CHALLENGER
The Final Voyage

The crew of *Challenger* 51-L *(left to right):* front row: Michael J. Smith, pilot; Francis R. (Dick) Scobee, commander; Ronald E. McNair, Mission Specialist 3. Rear row: Ellison S. Onizuka, Mission Specialist 1; S. Christa McAuliffe, Payload Specialist 1; Gregory B. Jarvis, Payload Specialist 2; Judith A. Resnik, Mission Specialist 2. (NASA)

CHALLENGER

The Final Voyage

RICHARD S. LEWIS

COLUMBIA UNIVERSITY PRESS   NEW YORK

Library of Congress Cataloging-in-Publication Data

Lewis, Richard S., 1916–
    Challenger : the final voyage / Richard S. Lewis.
        p.   cm.
    Includes index.
    ISBN 0-231-06490-X (alk. paper)
    1. Challenger (Spacecraft)—Accidents.   I. Title.
TL867.L49   1988   87-17929
363.1'24—dc19   CIP

Columbia University Press
New York   Guildford, Surrey

Copyright © 1988 Richard S. Lewis

All rights reserved

Printed in the United States of America

c 10 9 8 7 6 5 4 3 2

Clothbound editions of Columbia University Press are
Smyth-sewn and printed on permanent and durable
acid-free paper

This account of the accident that took their lives is dedicated to the crew of the *Challenger* 51-L mission.

Francis R. (Dick) Scobee, commander
Michael John Smith, pilot
Ellison S. Onizuka, mission specialist one
Judith Arlene Resnik, mission specialist two
Ronald Erwin McNair, mission specialist three
S. Christa McAuliffe, payload specialist one
Gregory Bruce Jarvis, payload specialist two

To strive, to seek, to find, and not to yield . . .
(Tennyson)

Contents

"We are at a watershed in NASA's history and the nation's space program. NASA's 28 year existence represents its infancy. We must use the knowledge and experience from this time to insure a strong future for NASA and the U.S. space program in the 21st century."

—Report of the U.S. House of Representatives Committee on Science and Technology. October 1986.

CHALLENGER
The Final Voyage

1. Seventy-Three Seconds

The morning was clear, bright, and cold, near freezing. Busses unloaded spouses, children, and parents of the shuttle crew at VIP observation sites. Three and one-half miles away, the space shuttle *Challenger* stood in silhouette against the morning light on launch pad 39B like some medieval battlement, exhaling white smoke. This image acquired a glow as the sun rose out of a cold, quiet sea. It was launch day at the Kennedy Space Center: January 28, 1986.

After five days of delay filled with wind, rain, and frustration, *Challenger* was ready to go on mission 51-L, the twenty-fifth for the world's first fleet of reusable manned spaceships and the tenth of the winged orbiter *Challenger*.[1] Public interest in the flight had been focused by a strong public-relations buildup on the first private citizen to fly aboard the space shuttle. She was Christa McAuliffe, 37, a high-school teacher who had been selected from among thousands of applicants to become the first teacher in space.

Mrs. McAuliffe's assignment was to demonstrate and explain the effects of microgravity (free fall in orbit) in the context of Newtonian physics and the scientific, commercial, and industrial applications of space flight. She would address an audience of schoolchildren via television from the spaceship.

The presence of this personable and attractive young woman was calculated to add a new dimension to the public's perception of the space program: space flight in America was no longer the exclusive province of astronaut test pilots, scientists, and engineers, but an experience to be shared by the whole society.

Now in its twenty-eighth year, the National Aeronautics and Space Administration was approaching a goal of quasi-airline operations in low Earth orbit. Flight schedules were devised on a yearly basis for scientific, commercial, and military payloads. With the space shuttle transportation system, the agency was preparing to establish a permanently crewed space station. Technology developed from that enterprise was expected to provide the means of constructing a base on the Moon for scientific and industrial development and of launching a manned expedition to Mars.

The shuttle was the basic unit of the space transportation infrastructure that would be required to realize these goals. After four test flights in 1981–82, the system was declared officially operational. *Columbia,* the first orbiter in the fleet, had demonstrated that despite minor problems the system was spaceworthy. *Challenger* was the second orbiter. It was followed by *Discovery* and *Atlantis*.

Challenger had been the orbiter most frequently flown. Now on the morning of its tenth launch, as it hung from its huge external propellant tank, flanked by the twin towers of its solid rocket boosters, *Challenger* represented the most advanced space transportation technology in the world. No other nation had achieved a reusable spaceship that could maneuver inside and outside the atmosphere and lift payloads with an Earth

---

[1] After 1983, shuttle missions were designated by the fiscal year in which they were originally scheduled, the launch site, and an alphabetical sequence. Thus 51-L referred to the 1985 fiscal year (5), the Kennedy Space Center (launch site 1) and the mission sequence.

Challenger stands on pad B ready for launch. (NASA)

weight of up to 32 tons. *Challenger* was the exemplar of American space leadership in the 1980s, as Apollo had been a decade earlier.

The morning of January 28, 1986 was the coldest on which NASA had ever attempted to launch a manned spacecraft. A cold front had swept down the Florida peninsula, ending high winds and threat of rain, and now the air was clear and relatively calm. Family member, relatives, and friends of the crew and a long list of Very Important Persons invited to observe the launch were bundled up in sweaters and jackets against a damp chill in the air.

The day was warming with sunrise. Through field glasses, ranks of icicles could be seen hanging from launch pad structures. Ice had built up overnight on beams and horizontal structures from water that had been allowed to trickle out of the pad water system to keep pipes from freezing. Most of the platform ice had been cleared away by the NASA pad inspection team. The team also removed ice that had formed on the surface of the pad's overpressure water troughs. The dangling icicles created a winter scene, bizarre in sub-tropical Florida.

Originally scheduled for launch December 23, 1985, the 51-L mission had been postponed four times and scrubbed once. Rain and high winds had forced delays of the previous mission, 61-C, a *Columbia* flight. It carried a congressional passenger, U.S. Representative Bill Nelson (D-Fla.), chairman of the House subcommittee on space science and applications. The 61-C delays—there were six—made it necessary to reschedule *Challenger* initially to January 23. The day before, NASA's Program Requirements Board postponed the launch to January 25 and then to January 26 because of the work backlog accumulated from the delay of the 61-C launch.

NASA officials invited hundreds of guests to see the launch of 51-L, a unique mission in which *Challenger* would serve as an orbital classroom. In addition to a group of 30 VIPs flown to Florida from Washington as NASA's special guests, hundreds more were invited to witness the launch from VIP observation sites and attend briefings. The guest list included educators, corporate sponsors of the Young Astronauts Council, members of the Michigan Republican party organization led by E. Spencer Abraham, chairman, members of the Teacher Astronaut Selection Panel, and a delegation from the People's Republic of China.

Prelaunch briefings were scheduled for these groups during the evening of January 25. The briefers included Dr. William R. Graham, acting NASA administrator; Dr. Lawrence F. Davenport, U.S. Assistant Secretary of Education; Barbara Morgan, the teacher-in-space backup for Christa McAuliffe; Richard G. Smith, director of the Kennedy Space Center; U.S. Senator E. J. (Jake) Garn (R-Utah), who had flown as a payload specialist on *Discovery* 51-D, April 12–19, 1985; and Vance D. Brand, commander of two previous missions, *Columbia* STS-5, November 11–16, 1982, and *Challenger* 41-B, February 3–11, 1984.

The January 26 launch was postponed by a forecast of unacceptable launch weather, and many of the special guests went home. The launch was reset to the morning of the 27th, and fueling began at 12:30 A.M. The crew was awakened at 5:07 A.M. and seated in the orbiter by 7:57. The countdown proceeded until it was halted at 9:10 A.M. when the ground crew reported a

problem with the exterior hatch handle. The problem was corrected at 10:30 A.M., but by this time high crosswinds had sprung up at the launch site and imperiled an emergency landing there in case the orbiter was forced to fly back by main engine failure. The high winds blowing across the three-mile runway of the Kennedy landing facility persisted, and the launch was scrubbed at 12:35 P.M.

Weather conditions allowing a safe emergency landing at Kennedy and other emergency landing sites were a requirement for launching the shuttle. During its 8½ minute ascent to orbit, the flight could be aborted only after the solid rocket boosters burned out and were dropped from the external tank. This took place at 2½ minutes. From then on, the orbiter was propelled to orbital altitude by its three main hydrogen-oxygen engines, which were fed from the big tank. If one of these engines failed, the flight could be aborted by separating the orbiter from the tank and steering it as a glider to an emergency landing strip.[2] The nearest, Kennedy Space Center, was considered safe if wind blowing across the runway did not exceed 10 knots and the skies were clear.

Depending on the velocity the orbiter reached before engine failure occurred, other emergency landing sites were available. There were two in Africa, one at Dakar, Senegal and the other at Casablanca, Morocco. Farther along the ascent track, emergency landings could be made in the Mojave Desert at Edwards Air Force Base, California or at White Sands, New Mexico.

The abort system was contingent on main engine failure. There was no abort system planned against failure of one or both solid rocket boosters during the first two minutes of the ascent. Nor was there any means of escape for the crew. Unlike Mercury, Gemini, and Apollo spacecraft, the orbiter was not equipped with a launch escape system during the solid rocket booster phase, the first stage of the ascent. Such a system had

[2] If more than one engine failed, an emergency landing was considered possible under some conditions. The alternative was ditching the orbiter in the sea.

Icicles on the mobile launch platform before launch. (NASA)

been considered during the development of the shuttle, but had been dropped, except for the temporary installation of aircraft-style ejection seats in *Columbia,* because failure of the solid rocket boosters after launch was considered highly improbable.

The Ice Problem

Following the scrub on the 27th, launch was reset to January 28. As the cold front passed through east central Florida, the winds were expected to die down, and the morning was predicted to be clear and cold. Temperatures were expected to fall to the low twenties overnight.

Mission managers assessed the possible effects of the cold weather on the launch, but the only threat they perceived to launch safety was ice on the pad structure and in the water troughs below. Ice breaking off the structure during launch could damage *Challenger's* heat shield, which consisted of 31,000 silicate-based tiles, all quite brittle. Extensive tile damage might expose the

More icicles hanging from the launch pad fixed service structure. *Challenger*'s left wing is visible in the background. (NASA)

aluminum skin of the orbiter to high heat during reentry into the atmosphere.

During the night, the accumulation of ice from dribbling water pipes on the pad structures caused concern at the launch control center. An "ice team" of engineers and technicians was sent to the pad at 1:35 A.M. on the 28th. Its report was not encouraging, but launch directors anticipated that most of the ice could be cleared away before sunrise or would melt shortly thereafter.

The Crew

The crew was awakened at 6:18 A.M., an hour later than scheduled because of a delay in the countdown caused by a fault in the ground liquid hydrogen tank fire detector. Following a somewhat leisurely breakfast, the crew received a weather briefing. There was some good and some bad news. The emergency landing site at Casablanca was declared "no-go" because of rain and a low ceiling, but weather was acceptable at Dakar, the prime transatlantic abort site. If both sites had been closed by weather, the launch would have been postponed again.

Ice conditions at the launch pad were described at the weather briefing, but if there was any concern about them, or about the effect of the unusually cold day on the performance of the shuttle at launch, it was not expressed to the crew.

The seven members arrived at the launch pad in the astronauts' van shortly after 8 A.M. and immediately rode up in the pad elevator to the white room where they entered the orbiter crew module. By 8:36 A.M., they were all strapped in their seats. Liftoff was to be at 9:38 A.M.

The ice team, meanwhile, had gone out to the pad a second time during the morning and had completed its inspection at 8:44 A.M. After hearing the team's report, the launch directorate decided to delay liftoff to allow additional time for the sun to melt ice on the pad. The ice was three inches thick on the pad's retractable service structure and covered the mobile launch platform, 160 feet long and 135 feet wide. Ice floated in the overpressure water troughs below the launch platform,[3] and icicles 6 to 12 inches long and five-eighths of an inch thick hung like stalactites from the 120- and 220-foot levels of the pad's fixed service structure. As the sun rose, the air temperature climbed from 26.1° to 36.2° Fahrenheit.

[3] Installed to prevent back pressure of rocket exhaust from moving the wing elevons, the body flap on the underside of the orbiter, and damaging struts.

Leaving the Operations and Checkout Building en route to the launch pad are Commander Francis R. (Dick) Scobee, Mission Specialists Judith A. Resnik and Ronald E. McNair; Pilot Michael J. Smith; Payload Specialist S. Christa McAuliffe; Mission Specialist Ellison Onizuka; Payload Specialist Gregory B. Jarvis. (NASA)

Four members of the crew were seated on the flight deck. By tradition, the forward left-hand seat was occupied by the commander, Francis R. (Dick) Scobee, 47, an Air Force–trained test pilot from Cle Elum, Washington. He had flown as pilot in *Challenger* on mission 41-C, April 6–13, 1984. The 51-L pilot in the right-hand seat was Michael J. Smith, 40, a Navy-trained test pilot from Beaufort, North Carolina. He was making his first space flight.

Although born and reared a continent apart, these two astronauts had strikingly similar physical attributes. NASA's public information dossiers confirmed that each was 6 feet tall and weighed 175 pounds. Each had brown hair and blue eyes. Both were graduates of the test pilot schools of their respective services. They appeared to express the Anglo-Saxon/North European phenotype that had been predominant in the Corps of Astronauts since its inception.

However, in the era of the space shuttle, that ethnic predominance gave way to an ethnic and gender mix that was more broadly representative of the American melting pot and that reflected the social policy of 1980s. In these respects, the crew of 51-L exemplified the diversity of American society more nearly than any other flight team.

The aft-center seat on the flight deck was occupied by Judith A. Resnik, Ph.D., 36, of Akron, Ohio, flight engineer. A mission specialist, she was a graduate in electrical engineering of Carnegie Mellon University, Pittsburgh, and had been awarded a doctorate in that field from the University of Maryland. She was the second American woman to fly in space, having logged 144 hours and 57 minutes in orbit on mission 41-D, the first flight of the orbiter *Discovery*, August 30–September 5, 1984. The aft right-hand seat was occupied by Air Force Lieutenant Colonel Ellison S. Onizuka,

Seated in the shuttle mission simulator in the positions they later occupied at launch are *(left to right)* Michael J. Smith, Ellison S. Onizuka, Judith A. Resnick, and Francis R. Scobee on the flight deck. (NASA)

Seated in the simulator middeck are S. Christa McAuliffe, Gregory B. Jarvis, and Ronald E. McNair. Standing left is the backup teacher in space, Barbara R. Morgan. At launch, Christa McAuliffe occupied the center seat and Greg Jarvis the right-hand seat. (NASA)

39, mission specialist from Kealakekua, Kona, Hawaii, an aerospace engineer with a master's degree. He had served as a flight engineering instructor at the Air Force Test Pilot School, and had taken part in checking out the first two orbital test flights of the shuttle. He was the first American of Japanese descent to become a member of a flight crew.

On *Challenger*'s middeck, the left-hand seat was occupied by Ronald E. McNair, Ph.D., 35, mission specialist from Lake City, South Carolina. He was a physicist with a bachelor of science degree from North Carolina A & M University and a doctorate from the Massachusetts Institute of Technology. He had become the second black astronaut to fly in space as a member of the crew of *Challenger* 41-B, February 3–11, 1984. This was his second mission.

The middeck right-hand seat was occupied by Gregory B. Jarvis, 41, a payload specialist from Detroit. He was assigned to the mission as a technical expert from the Hughes Aircraft Company to operate experiments on the behavior of fluids in microgravity (the so-called zero g). The results were expected to be useful in the design of advanced liquid propulsion systems. Hughes named its new multiple satellite launcher for him.

The middeck center seat was occupied by NASA's teacher in space, Sharon Christa Corrigan McAuliffe, a native of Boston with a master's degree in education. The Teacher-in-Space Program had been proposed by President Reagan who proclaimed it a means of communicating to the nation's schoolchildren (and to their parents) the nature and potential value of the new frontier the American space establishment was exploring.

Mrs. McAuliffe was a history and social studies teacher at the Concord High School, Concord, New Hampshire. She had been selected by NASA on July 19, 1985 and had undergone training for the flight at the Johnson Space Center, Houston. Later, a journalist was to be selected to fly a shuttle mission as an observer and narrator in the White House program of bringing space to the people.

The program of the teacher in space consisted of two classroom lessons to be televised from *Challenger* on flight day 6. The first, titled "The Ultimate Field Trip," was to describe life aboard a spacecraft in orbit. The second was to explain methods of exploring space and describe the prospects of manufacturing new products in space, where tests had demonstrated that a microgravity environment increased the purity of pharmaceuticals and the precision of microelectronic devices manufactured there.

Here was the crew of *Challenger* 51-L on the frosty morning of January 28, 1986 as the countdown approached liftoff, ready to embark on the first teaching mission of the space age.

The Cargo

The orbiter is the largest space transport ever built. Its cargo bay was designed to carry 65,000 pounds of payload into low Earth orbit, but cargoes have been considerably less than that, partly because of constraint on the weight at landing. *Challenger*'s total payload on 51-L weighed 48,361 pounds.

In the cargo bay, *Challenger* carried TDRS-B, the second of three powerful tracking and data-relay communications satellites. They were designed to relay all NASA satellite communications to a central ground station at White Sands, New Mexico, for distribution over conventional telephone circuits. When completed, the TDRS system would replace most of NASA's ground stations and would provide full orbital coverage of

shuttle-ground communications. The TDRS satellites, like the commercial comsats, would operate in geostationary orbit, 22,300 miles above the equator, where a satellite appears to hover in a fixed position. The first of the series, TDRS-1, was delivered to low Earth orbit by *Challenger* in 1983 and was maneuvered into its operational orbit by auxiliary thrusters after its upper-stage booster failed.

Also in the cargo bay was a 2,250-pound observatory called Spartan-Halley, which had been built to photograph Halley's comet in ultraviolet light. It was to be lifted out of the cargo bay by the orbiter's remote manipulator arm and released over the side in temporary free flight. Tagging along with the ship, the automated observatory would make spectrographic and photographic observations for 22 orbits before being recovered by the arm and restowed in the cargo bay.

The fluid-dynamics experiments that were to be conducted by Gregory Jarvis were carried in middeck. The equipment consisted of small, transparent tanks in which the behavior of fluids could be observed during orbital flight. Fluid motions were to be recorded by video cameras as a tank was set spinning at 10 revolutions per minute and as tanks were filled at several levels. The interaction of fluids with motions of the orbiter was to be observed. These experiments were relevant to the development of propulsion systems for orbital transfer vehicles, contemplated for moving cargoes between a space station and geostationary or lunar orbits.

The TDRS, the Halley's comet observatory, and the fluid-dynamics experiments were of prime importance to space science and technology, but in terms of public interest, the main event of the flight would be the lessons by the teacher in space. This was an event with which millions of Americans of all ages could readily identify. The cynosure of this flight was the New England schoolteacher, a role model for young women in her profession as well as a mother of two children. Her children, her husband, and her parents were on hand with the families of her crewmates to watch the liftoff of this magnificent vehicle with pride and excitement. Mission 51-L carried a message that space was for everyone.

The ice team went out on its third inspection at 10:30 A.M., removed more ice from the deck of the mobile launch platform, and pulled more floating ice out of the water troughs with fishnets. Its measurements showed that the temperature of the left booster was 33° F while it was 19° F on the right booster. No significance was attached to the discrepancy at the time. The difference was attributed to wind blowing off the cryogenic external tank onto the right booster.

Final Countdown—Crew

As the five men and two women of the crew rode up the pad elevator and entered the orbiter, their banter displayed high spirits and the humor of a well-integrated team. The rain had stopped, and the sun was shining. One by one they entered the crew module and reclined in their seats, assisted by Manley L. (Sonny) Carter, Jr., the astronaut support person (ASP). Each then followed a ritual of opening communications with the orbiter test conductor (OTC), Roberta Wyrick, and the Lockheed test director (LTD), Ken Jenicek.

From the shuttle landing facility came an announcement that a NASA T-38 jet had taken off to look at the weather and winds aloft.

Sonny Carter addressed Commander Dick Scobee in the left-hand seat. "Ready for com [communications] check?"

Scobee: Uh huh. Good morning, Roberta.

The commander then addressed the LTD, Jenicek:

Scobee: Good morning. [Then responding to Jenicek, he added] Loud and clear.[4]

[4] Only the voices of the crew and astronaut support person were fully recorded at Mission Control, although the recording did carry brief responses by test personnel.

Radio communication checks were conducted through the helmet, visor down.

Scobee then greeted Pilot Mike Smith as he took his seat.

Carter: Okay, Mike. You oughta be able to hear me.
Smith: Okay. Hear you loud and clear. My helmet's good.

There followed a check of shuttle systems, with Scobee and Smith discussing the results with test personnel. Lying nearly supine in their seats, they looked out at a bright blue sky and beyond it, a hazy sea.

Smith: These clouds look like they're going away from us up here.
Scobee: I think they are. Looks good over this way.
Smith: Wow! Boy! The sun feels good this morning.
Carter: You should have been here at 2 A.M.
Scobee: Ice skating on the MLP [mobile launch platform]? You guys up here working?
Carter: It's a lot of fun.
Scobee: Winds out of the north blowing pretty strong, looks like?
Smith: They are or aren't?
Scobee: They're not. The anemometer isn't going around, and the arrow's pointed in the right direction.
Carter: Say, OTC [orbiter test conductor]. MS-1, com check. [Mission Specialist 1 was Ellison Onizuka.]
Onizuka: Read you loud and clear.
OTC(Wyrick): Good morning, Ellison.
Onizuka: Good morning.
Carter: Visor down and talk to LTD [Lockheed Test Director, Jenicek].
Onizuka: LTD, MS-1 com check. You're loud and clear. Good morning.
Carter: You know, I think these visors are cold from being outside and are fogging up. Here's the pocket checklist.
Onizuka: Okay, thank you. See you later on. Kind of cold, this morning.
Smith: Up here, Ellison, the sun's shining in. At least we've got the crew arranged right for people who like the warm and the cool. [Commander and pilot

were seated forward on the flight deck with the two mission specialists, Onizuka and Judith Resnik (MS-2), behind them.] Got you out of the sun.

Scobee: You got that right.

Onizuka: My nose is freezing.

Carter: Good morning, Judy.

Resnik: Cowabunga.

Carter: Heyyy!

Scobee: Loud and clear, there, Judy.

Carter: OTC.

Resnik: OTC, MS-2 radio check. Good morning.

Carter: Visor up there, if you like.

Resnik: It feels like you know in the T-38 when your exhalation valve sticks, 'cause I'm not tightened down enough.

Carter: Roger. But I didn't hear a leak.

Resnik: No, I didn't hear a leak either. I exhaled. I heard a pop. Am I okay up here?

Carter [assisting her]: There you go. Let me go over these panels one more time to make sure circuit breakers and switches are all okay. [He then turns to Christa McAuliffe, Payload Specialist 1 (PS-1)]: Okay, Christa, you oughta be able to hear me.

McAuliffe: Real fine.

Carter: Okay, real good. Put your other arm through there and I'll hold it for you. Okay. Talk to the OTC on that button.

McAuliffe: OTC, PS-1.

OTC: Loud and clear.

McAuliffe: Good morning— I hope so, too.

Carter: Now, just put your visor down on the right.

McAuliffe: It's down.

Carter: Now tighten your helmet a little bit on the back. Make sure it's snug but not too tight, though, and then push this button right here and tell the LTD com check.

McAuliffe: LTD, PS-1. Com check.

Jenicek: Have you loud and clear.

McAuliffe: Good morning.

Carter: Okay. Now raise your visor with a little push with your right hand. Right hand here. It'll fog a little; those things are cold, and we're a little warm. Now, we'll put that beauty [emergency air pack] where it feels most comfortable for you, about there

[beside the seat]. Feel where that is now. Okay? Okay, you're ready. Doin' good. Watch your arm there, Christa. There you go. Christa, while I hook up Greg [Jarvis], you're gonna lose com for a while. Good morning, Greg.

Jarvis [PS-2]: Good morning, Billy Bob. How are you?

Carter: Fine. And you?

Jarvis: Fannnnntastic!

Carter: Tell the OTC on that other button you're the PS-2.

Jarvis: OTC, PS-2 radio check.

OTC: Copy.

Jarvis: Roberta, how are you doing this morning?

Unidentified: NASA's not the pumpkin today.

[Laughter.]

Carter: Now with your right hand, lower your visor, tighten your helmet, not too tight, talk to the LTD, and tell him you're PS-2.

Jarvis: LTD, this is PS-2 radio check. Good morning.

Carter [to Ronald McNair]: Good morning, Ron. How you doing?

McNair [MS-3]: Morning. Okay.

Carter: Okay, Ron. Tell the OTC MS-3 radio check.

McNair: OTC, MS-3 radio check.

OTC: Read you loud and clear, good morning.

Carter: Dick, we're gonna take the exhalation valve out of Ron's helmet and look at it and put it back in.

Scobee: Okay.

Carter: We didn't change the exhalation valve.

Following a rapid series of communication checks with test personnel, Carter announced that he was going to terminate the communications tests and get out of the cockpit, where he had been helping crew members adjust themselves in their seats.

Scobee: Commander understands a T equals zero no earlier than 11:08. [That is, launch no earlier than 11:08 A.M.] Everybody hear that?

Jarvis: Unfortunately.

The crew members were beginning to feel the strain of lying supine in their couches in one position.

Smith: I feel like I'm at the four-hour point of yesterday.

Scobee: Yeah.
Resnik: I feel like I'm past it. My butt is dead already. . . . Okay, Ellison, get out of there.
[Laughter.]
Onizuka: That's too low.
[Laughter.]
Smith: Crew [garble] . . . crew gynecologist.
[Laughter.]
Jarvis: We need a round of applause for the close-out people.
Resnik: Is that snow?
Scobee: Yep, that's snow. [Frost was blowing off the external tank.]
McNair: You're kidding. You see snow on the window?
Scobee: Yep, it's clouded up out here and started to snow.
Jarvis: Yeah, fat chance.
[Laughter.]
Scobee: Blue skies.
Onizuka: Coming right off the ET [external tank].
Scobee: It seems to be coming from over off the tower somewhere.
Resnik: All those trickles that they're trickling; 100 gallons a minute are adding up. [She refers to water being allowed to trickle out of the pad water system to prevent the pipes from freezing.]
Onizuka: Where are they getting all that water?
Resnik: Your tax dollars.
Scobee: Yeah, that's probably special grade water.

The crew then tested and made adjustments to cabin pressure, which was reading 15.1 pounds per square inch, slightly higher than ambient atmospheric pressure outside at 14.7 psi. The cabin seal was tight.

Onizuka: Ice team out here?
Scobee: Don't see anybody out on top. Must be around the base. Ice pickin'.
[Laughter.]
Onizuka: 10:38 launch?
Resnik: That's indefinite.
Onizuka: How about half an hour?
Resnik: It's half an hour from now but it's still indefinite.
Scobee: We should've slept an extra hour this morning.

Jarvis: They're probably making a fortune selling coffee and doughnuts at the viewing areas.
Scobee: How about that. We should have gotten some.
Resnik: A few hot toddies.
Scobee: Space looks great . . . soakin' up a little bit.
McAuliffe: It'll be cold out there today.
Scobee: Could be hot.
Resnik: Oh, that's okay. My bun is dying.
Jarvis: Ellison and I'll massage it for you. . . .
Scobee: Hah!
Resnik: Ellison's not even interested.
Jarvis: I'll bet you could wake him up for that.
Resnik: Hah!
Scobee: Maybe.

The dialogue suddenly switched to the business at hand. And then switched back.

Scobee: CDR [commander] go. [There followed a report on cabin pressure.]
Resnik: 15.2 [pounds per square inch].
Scobee: Cabin press is 15.2 up here.
Resnik: After sitting like this, I can't imagine how anybody can hang in those gravity boots.
Scobee: They don't do it for four hours!
Resnik: Some people do.
Scobee: Oh, they do?
Resnik: For a long time.
Scobee: There's no pressure points other than your ankles, I guess.
[Laughter.]
Jarvis: They only recommend it for about 4 or 5 minutes.
Resnik: I saw one guy once do about a hundred situps from vertical all the way up to like you'd normally do a situp, but vertical.
Jarvis: That takes some very, very strong abdominal muscles, I'll tell you that.
Resnik: Takes some pretty weak, weak brains.
Launch Control Center: We're planning to come out of this hold on time.
Scobee: All riiight. Roger, go ahead. That's great.
Smith: All right.
Scobee: We're doing fine, just fine.
Resnik: One 10 . . . minute hold so that's one 30 for

11:38. That's the time they talked about in the quarters last night.

Scobee: Sure was.

Resnik: An hour and a half, you don't even have your seat comfortable yet.

Scobee: Oh, I know.

Resnik: Remember my foot there. I can move up.

Scobee: I'm sorry, I'll get out of the way. Everything straight back there, J.R.? [Nickname for Judith Resnik.]

Resnik: Yeah.

Scobee: There goes a sea gull.

Resnik: He better get out of the way.

Scobee: He'll be a long way from here by the time we launch.

Resnik: He's built a nest by now.

Scobee: A bunch of birds' nests in the box up here would be real surprised in about an hour. [He refers to the fact that this was the first launch from pad B since the Apollo era.]

Resnik: What year was this used last?

Scobee: Long time ago.

Smith: Midseventies.

Resnik: What came off here?

Scobee: About ten years ago.

Resnik [back to business again]: Mike, your boiler preact is next. [Preliminary to starting up the auxiliary power units.]

Scobee: Let's get 'em [the boilers] started.

Smith: OTC, pilot. H_2O boiler preact is complete.

Scobee: An hour and counting.

Smith: Wind sure isn't as high as it was, looking at the waves out here.

Scobee: The anemometer is just a little bit out of the north.

Smith: Ellison, I'm getting hungry.

Resnik: God, I had two steaks.

Scobee: Oughta hold you for two days.

Resnik: Probably keep me off the pot anyway.

Onizuka: Steaks are good.

Smith: A shot of cholesterol, eh? Is your potassium level up so you can launch?

Resnik: Did you eat a banana this morning?

Scobee: OTC, CDR. Pass BFS transfer is complete. [He refers to the backup flight software for the flight computers.] Gaseous nitrogen pressure is complete.

Smith then recited fuel and oxidizer tank readings for the fuel cell batteries. There was no change from the day before when the mission was scrubbed.

Scobee: Another nine minutes there, gang.

Resnik: My butt's gonna like zero g a lot better than these seats.

Scobee: Mine too. Well, y'all on the middeck, it's clear blue out today.

Smith: May get by T minus 9 today. [A crucial point in the countdown.]

Resnik: Get past T minus zero.

Smith: J.R., you're like Scobee, you want. . . .

Resnik: We're on internal power anyway, so are the payload buses on?

Smith: Gotcha.

Resnik: Oh, Scobes [the nickname she sometimes used for Scobee]. You may or may not get a dp/dt [change in pressure with time], but you might also get the high-press alarm as the cabin expands when we go up.

Scobee: Okay, Judy. . . . Seven minutes. There goes the arm [the crew access arm].

Smith: There it goes.

Resnik: Bye, bye. Tighten your straps.

Smith: Okay, APU [auxiliary power unit] controller power three to on. Fuel tank valves are coming open.

Onizuka: Elevon movement. [Movable wing surface.]

Smith: APUs are coming up.

Scobee: Go pressure on all three APUs. All three up.

Smith [to OTC]: We've got three good APUs.

Scobee: Visors are coming down [in response to launch control command to close helmet visors].

McAuliffe and Resnik [testing]: Can you hear me, Scobes?

Scobee: Yeah, loud and clear.

Resnik: Okay.

Scobee: Can you hear okay?

Resnik: Yeah.

Smith [to OTC]: Caution and warning is clear and no unexpected messages.

Scobee: Welcome to space, guys. Two minutes downstairs.

Final Countdown—Public

The two-hour hold in the countdown during which the crew had been waiting restlessly for some sign that at last they might fly this day came to an end, and the crew was relieved that the long wait might not have been in vain. At the viewing sites, their families, friends, and thousands of other spectators were alerted to this juncture in the countdown by the voice of Hugh Harris, the Kennedy Space Center chief of public information, booming out of loudspeakers from the firing room. It was 11:28 A.M.

"One minute away from picking up the count for the final nine minutes in the countdown," Harris announced. "The countdown is simply a series of checks that people go through to ensure that everything is ready for flight. The countdown for a launch like 51-L is four volumes and more than 2,000 pages."

Scraps of dialogue between the *Challenger* crew and launch control filtered through public address system, sounding impressively technical and bordering on the arcane. Hugh Harris' commentary sought to explain what was going on in the complex process of launching a spaceship.

The commentary related: "T minus 8 minutes 30 seconds and counting. All the flight recorders are turned on. Mission Control [at Houston] has turned on the auxiliary data system. The package of flight data from the aerodynamic information [system] comes back as the orbiter flies through the atmosphere.

"Coming up on the eight-minutes point. T minus 8 minutes and counting. Orbiter test conductor Roberta Wyrick has requested that Houston send the stored program commands, which is the final update on antenna management based on liftoff time and sets the system which makes the orbiter compatible with downrange tracking stations."

From *Challenger* a voice said, "Okay, that's the point."

The commentary continued: "T minus 7 minutes, 30 seconds, and the ground launch sequencer has started retracting the orbiter crew access arm. This is the walkway used by astronauts to climb into the vehicle. And that arm can be put back in 15 to 20 seconds if an emergency should arise. Coming up on the seven-minute point in the countdown. T minus 7 minutes and counting. The next major step will be when Pilot Mike Smith is given a 'go' to perform the auxiliary power unit prestart.[5] T minus 6 minutes and 50 seconds and counting."

Launch control called for voice recorders. "Roger, wilco," responded *Challenger*.

"T minus 6 minutes and the orbiter test conductor has given pilot Mike Smith the 'go' to perform the auxiliary power unit prestart. He has reported back that prestart is complete. T minus 5 minutes and 30 seconds and counting, and Mission Control has transmitted the signal to start the on-board flight recorders. The two recorders will collect measurements of the shuttle system performance during flight to be played back after the mission. Coming up on the five-minute point. This is a major milestone where we go for auxiliary power unit start. T minus 5 minutes."

Launch control: Let's go for orbiter APU start.
Challenger: . . . performed APU start.

Commentary: "We heard the pilot ordered to perform APU start. Lox [liquid oxygen] replenish has been terminated, and liquid oxygen drainback has been initiated. Pilot Mike Smith now flipping the three switches in the cockpit to start each of the three auxiliary power units. T minus 4 minutes 30 seconds and counting.

"The solid rocket booster and external safe and arm devices have been armed. We have had a report

---

[5] The auxiliary power units (APUs) provide hydraulic pressure to move the orbiter's aero surfaces and engine nozzles for steering. They also assist deployment of the landing gear and application of the brakes.

back from Mike Smith that we have three good auxiliary power units. Main fuel-valve heaters on the shuttle main engines have been turned on in preparation for engine start. T minus 4 minutes and counting. The flight crew has been reminded to close their airtight visors on their launch and entry helmets. And a final purge sequence of main engines is under way.

"T minus 3 minutes and 45 seconds and counting. The orbiter aerosurface test is started. The flight control surfaces are being moved in a preprogrammed pattern to verify that they are ready for launch. Orbiter ground support equipment power bus has been turned off and the vehicle is now on internal power."

The Crew at Launch

T minus 1:47.   **Smith:** Okay, there goes the lox arm [A vent arm covering the top of the external tank during liquid-oxygen loading].

T minus 1:46.   **Scobee:** Goes the beanie cap [tip of the arm].

T minus 1:44.   **Onizuka:** Doesn't it go the other way? [Laughter.]

T minus 1:39.   **Smith:** God, I hope not, Ellison.

T minus 1:38.   **Onizuka:** I couldn't see it moving. It was behind the center screen.

T minus 1:33.   **Resnik:** Got your harnesses locked?

T minus 1:29.   **Smith:** What for?

T minus 1:28.   **Scobee:** I won't lock mine. I might have to reach something.

T minus 1:24.   **Smith:** Oooh, kaaay.

T minus 1:04.   **Onizuka:** Dick's thinking of somebody there.

T minus 1:03.   **Scobee:** Uh, huh.

T minus 0:59.   **Scobee:** One minute downstairs.

T minus 0:52.   **Resnik:** Cabin pressure is probably going to give us an alarm.

T minus 0:50.   **Scobee:** Okay.

T minus 0:47.   **Scobee:** Okay, there.

T minus 0:43.   **Smith:** Alarm looks good.

T minus 0:42.   **Scobee:** Okay.

T minus 0:40.   **Smith:** Ullage pressures are up [tank propellant pressure].

T minus 0:34.   **Smith:** Right engine helium tank is just a little bit low.

T minus 0:32.   **Scobee:** It was yesterday, too. [This dialogue refers to helium pressurization in orbiter main engine number 2.]

T minus 0:31.   **Smith:** Okay.

T minus 0:30.   **Scobee:** Thirty seconds down there.

T minus 0:25.   **Smith:** Remember the red button when you make a roll call. [To assure communications when Scobee announced the shuttle was rolled to ascent attitude.]

T minus 0:23.   **Scobee:** I won't do that. Thanks a lot.

T minus 0:15.   **Scobee:** Fifteen.

T minus 0:06.   **Scobee:** There they go, guys. [*Challenger*'s main engines begin firing.]

T minus 0:06.   **Resnik:** All right.

T minus 0:06.   **Scobee:** Three at a hundred. [All three orbiter main engines operating at 100 percent of rated power.]

Challenger's two solid fuel boosters ignited. With a groundshaking roar and vast billows of white smoke, *Challenger* rose on pillars of orange-yellow fire. Observers watched in awe and then broke into cheers, screams, applause as the spaceship, the product of a thousand years of technical evolution, thundered majestically into the clear sky. The final flight of *Challenger* began at 11:38:00:010 Eastern Standard Time, January 28, 1986.

T minus 0:00.   **Resnik:** Aaäll right!

T plus 0:01.   **Smith:** Here we go!

"Liftoff!" cried Hugh Harris at Kennedy launch control. "Liftoff of the twenty-fifth space shuttle mission and it has cleared the tower!"

At this point, control of the launch shifted from Florida to Texas as the Mission Control Center (MCC), Houston took over the flight. The commentary was taken up there by Stephen Nesbitt, public affairs officer (PAO) at the Johnson Space Center. Mission Control could be heard talking to *Challenger* through the public address system at Kennedy Space Center.[6]

    MCC: Watch your roll, *Challenger.*

T plus 0:07. **Scobee:** Houston, *Challenger* roll program. [The first maneuver after liftoff was to rotate the shuttle to ascent attitude, in which the orbiter was turned to face the Earth.]

    Houston PAO: . . . roll program confirmed. *Challenger* now heading down range. Engines beginning to throttle down to 94 percent. Normal throttle for most of the flight is 104 percent. Will throttle down to 65 percent shortly.[7] Three engines running normally. Three good fuel cells [electric power generators]. Three good APUs. Velocity 2,257 feet per second [1,538 miles per hour]. Altitude 4.3 nautical miles. Down-range distance, 3 nautical miles. Engines throttling up. Three engines now at 104 percent.

T plus 0:11. **Smith:** Go, you mother.

T plus 0:14. **Resnik:** LVLH. [A reminder about switch configurations—local vertical, local horizontal.]

T plus 0:15. **Resnik:** [one word deleted by NASA] hot!

T plus 0:16. **Scobee:** Oooh kaay.

T plus 0:19. **Smith:** Looks like we've got a lot of wind up here today.

T plus 0:20. **Scobee:** Yeah.

T plus 0:22. **Scobee:** It's a little hard to see out my window here.

T plus 0:28. **Smith:** There's 10,000 feet and mach point five.

T plus 0:35. **Scobee:** Point nine.

T plus 0:40. **Smith:** There's mach one.

T plus 0:41. **Scobee:** Going through 19,000 feet.

T plus 0:43. **Scobee:** Okay, we're throttling down.

T plus 0:57. **Scobee:** Throttling up.

T plus 0:58. **Smith:** Throttle up.

T plus 0:59. **Scobee:** Roger.

T plus 1:02. **Smith:** Thirty-five thousand. Going through one point five.

T plus 1:05. **Scobee:** Reading 486 on mine. [Airspeed indicator check.]

T plus 1:07. **Smith:** Yep. That's what I've got, too.

    MCC, Houston: Go at throttle up. [Continue at full throttle.]

T plus 1:10. **Scobee:** Roger, go at throttle up.

Racing skyward at 2,900 feet per second, *Challenger* had reached an altitude of 50,800 feet and was 7 nautical miles down range from the launch site when camera 207 on the ground saw a brilliant glow on one side of the external tank.[8] The glow blossomed into orange-yellow flame and in seconds grew to a gigantic fireball with a nimbus of gray-white smoke that formed a multi-lobed cumulus cloud filled with streaks of fire. The cloud encompassed the shuttle, but for a moment or two the exhausts of the two solid rocket boosters could be seen through it, together with exhaust plumes from the orbiter's main engines. Vaportrails from these engines streamed behind the cloud like tattered banners.

T plus 1:13. **Smith:** Uhh . . . oh!

It was the last recorded utterance from the crew as the orbiter broke up. Houston did not hear it.

    From the viewpoint of observers on the ground, a nightmare took form in the sky, visible for miles up and down the Atlantic coast. The twin solid rocket boosters began flying away from the moving cumulus of smoke

---

[6] The Mission Control communicator is usually an astronaut who speaks only to the crew. The public affairs commentator speaks to the news media. Sometimes both can be heard over the NASA communications network. Crew chatter on the orbiter intercom is not broadcast but is recorded at Mission Control.

[7] Power is reduced to slow ascent as the shuttle approaches the region of maximum dynamic pressure (Max Q) about one minute after liftoff.

[8] *Report of the Presidential Commission on the Space Shuttle Challenger Accident* (Washington, D.C., 1986) vol. 1, p. 26; vol. 2, p. L-56; vol. 3, p. O-158; Commentary, Johnson Space Center, Houston, Tex.

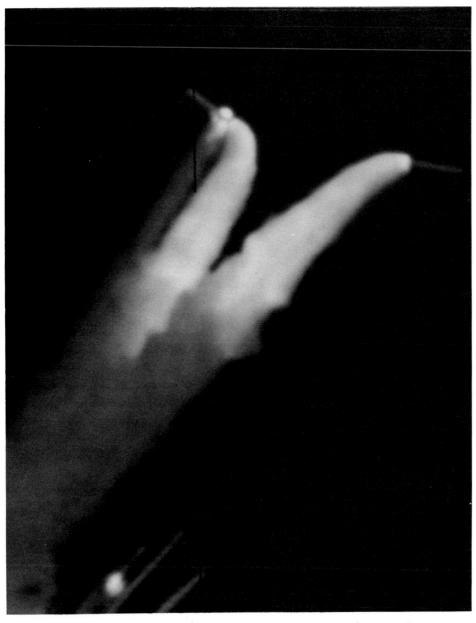

Breaking away from the flaming external tank, the solid rocket boosters fly off in opposite directions at 76.4 seconds after launch. (NASA)

and fire, diverging in fiery arcs. Screams of horror rose from thousands of watchers. Families of the crew looked at the scene in disbelief, unable at first to comprehend what they were seeing. Beyond the spreading cloud of darkening smoke, the sky was filled by debris falling into the ocean from 104,000 feet (19.7 miles), where radar showed the shuttle breaking up, to 122,400 feet (28.19 miles) peak altitude of debris.

Suddenly, the solid rocket boosters disintegrated.

The Air Force range safety officer detonated their "destruct" packages of explosives by radio signal at 101,300 feet (19.13 miles). One of them, partly filled with propellant burning at 5,600° F, had been heading for the coastal community of New Smyrna Beach.[9]

The external tank collapsed like a torn balloon and was reduced to wreckage by its flaming propellant. The

[9]Altitudes derived from radar data, *Commission Report*, vol. 3, p. O-158.

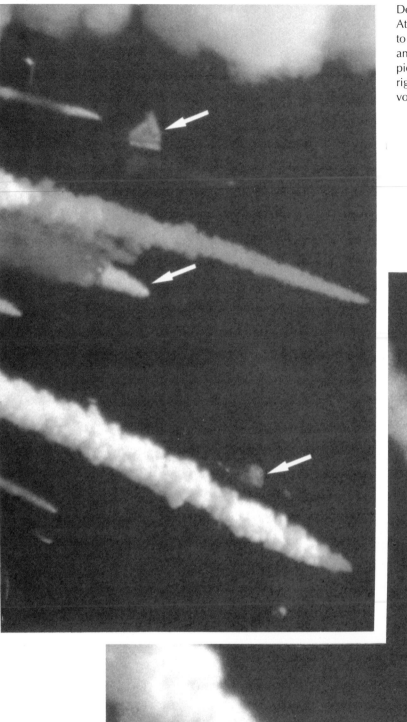

Debris hurtles out of the fireball at 78 seconds. At left, top arrow points to left wing, center arrow to main engines still burning residual propellant, and bottom arrow to forward fuselage. At right, piece identified as crew cabin falls free, lower right. (NASA—montage from *Commission Report*, vol. 1)

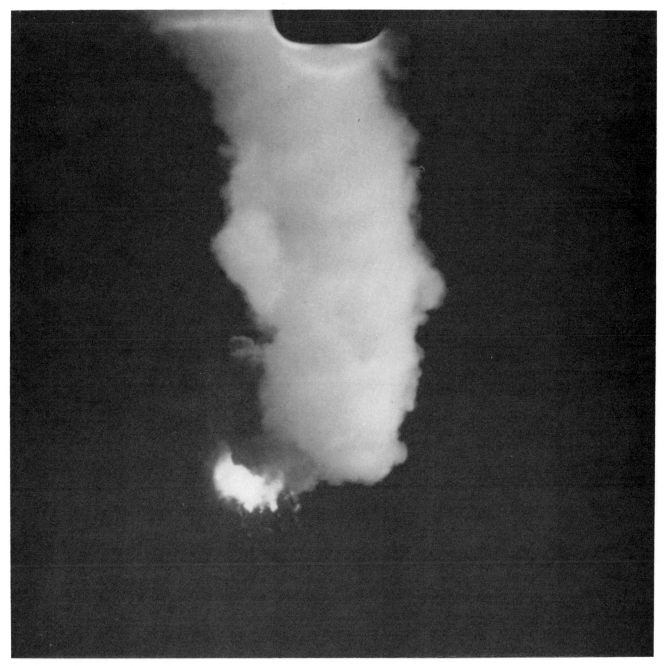

The right booster is destroyed by the range safety officer at 109.6 seconds. This photo was logged at 111 seconds. (NASA)

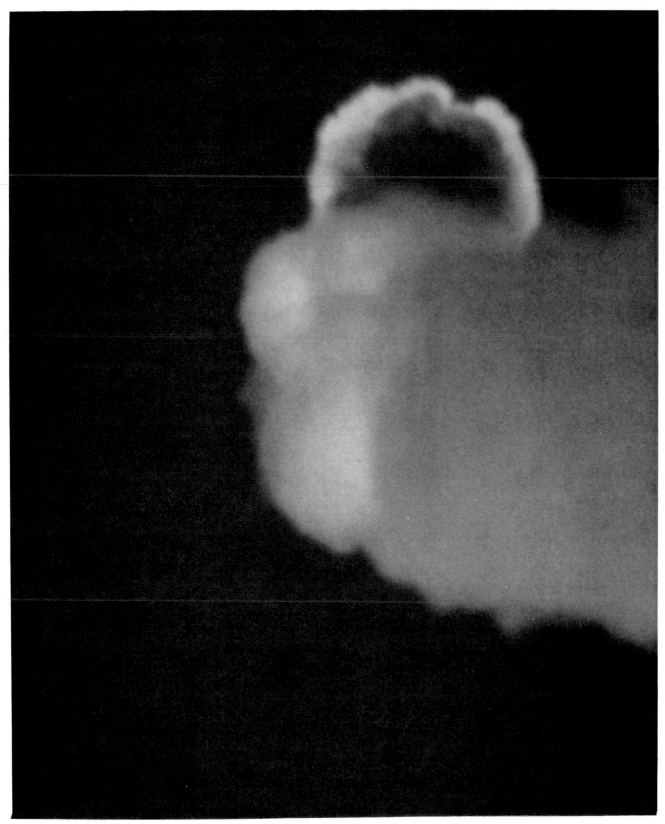

Telescopic camera zeroes in on the fireball of the right-hand solid rocket booster explosion.

orbiter was nowhere to be seen. It had vanished with the crew into the fireball. Beyond the pall of smoke and descending wreckage a lone parachute appeared, drifting casually down toward the sea. It was a 54-foot drogue in the parachute descent system of one booster, a system designed to allow recovery of the boosters at sea. Highly visible against the deep blue of the sky, the parachute, still attached to the forward section, or frustrum, of the booster, led hundreds of watchers to hope that the crew was drifting down to safety and would soon be rescued. The lack of a launch escape system on the orbiter was not popularly known, and some of the radio and television commentators who were describing the scene seemed unaware that such a system did not exist.

The white parachute went into the sea, like the flag of a sinking ship. From Mission Control came silence. Then the commentary was resumed in a tense monotone.

PAO: Flight controllers are looking very carefully at the situation. Obviously a major malfunction. We have no downlink. We have a report from the flight dynamics officer that the vehicle has exploded. The flight director confirms that. We are looking at checking with the recovery forces to see what can be done at this point. Contingency procedures are in effect. We will report more as we have information available. Again, a repeat, we have a report relayed through the flight dynamics officer that the vehicle has exploded. We are looking at all contingency operations, waiting for word of any recovery forces in the down-range field.

As Houston struggled to define the situation, a belief persisted among observers and even some NASA personnel that somehow the crew might have survived the holocaust and would be drifting in the insulated, airtight crew module. It was supposed, briefly, that it was possible for the crew to detach the orbiter from the external tank and booster and somehow survive a fall

at first estimated at 9 miles into the ocean.[10] The vision of the drogue parachute had lent a few moments of hope to that supposition. But veteran observers in the press corps and NASA people for the most part were certain that all had been lost. The space agency and the nation now observed the consequences of the 1972 decision to allow a manned spacecraft to be designed and flown without a launch escape system.

In the shuttle design, the crew was committed to flight during the first two minutes of the ascent to orbit, when the boosters were firing. Only after that first stage did escape become possible through the flight abort and contingency landing system described earlier. During its first two minutes of flight, the orbiter was a death trap for the crew if a booster failed. That fact had not been publicized.

When the commentary from Houston resumed, the public affairs officer stated that the vehicle, apparently having exploded, struck the water at a point 28.64° north latitude and 80.28° west longitude.

"We are awaiting verification from— as to the location of the recovery forces in the field to see what may be possible at this point. We will keep you advised as further information becomes available. This is Mission Control."

After a pause, the commentary resumed. "This is Mission Control. We are coordinating with recovery forces in the field. Range safety equipment, recovery vehicles intended for the recovery of the solid rocket boosters, are in the general area. To repeat: We had an apparently normal ascent with the data coming from all positions up through approximately the time of main engine throttle back up to 104 percent. At about approximately a minute or so into the flight, there was an apparent explosion. The flight dynamics officer reported that tracking reported that the vehicle had exploded and impacted into the water. . . . Recovery forces are pro-

[10] The initial estimate was based on the first sign of the "major malfunction" at 73.2 seconds. At this time, *Challenger* had reached an altitude of 9.63 miles (50,800 feet), according to radar data.

ceeding to the area including ships and C-130 aircraft. Flight controllers are reviewing their data here in Mission Control. We will provide you with more information as it becomes available. This is Mission Control, Houston.''

Nesbitt continued to tell as much as he was told by controllers, refraining from injecting speculation into the commentary. His was the most excruciating task in the annals of NASA public information. At NASA headquarters and the Kennedy and Johnson space centers, the public affairs people were under enormous pressure to explain the tragedy, but they knew little more than the fact that *Challenger* and its crew had vanished in an incredible disaster 73 seconds after liftoff.

Reaction

At the VIP observation sites, family members, relatives, and friends of the crew lapsed from disbelief into shock as they realized they had seen a son, a daughter, a wife, a husband vanish in the worst disaster of the space age.

Crew family members who had been waiting for the liftoff all morning joined hundreds of others at the VIP and press sites in applause and cheers as *Challenger* cleared the tower and ascended over the ocean. Launch of a manned spacecraft is certainly one of the most thrilling spectacles of the twentieth century. Then horror blossomed in the sky as the boosters tore away from the orbiter and the tank a minute too early and the fireball filled the sky.

Many family members were watching from the roof of the launch control center. There were June Scobee, the commander's wife, his son, Richard W., 21, soon to be graduated from the Air Force Academy, and daughter, Mrs. Kathie R. Krause, and her husband, Justin; Jane Smith, wife of the pilot, and their children, Scott, 17, Alison, 14, and Erin, 9; Lorna Onizuka, wife of Ellison Onizuka, and their children Janelle, 16, and Darien, 10; Cheryl McNair, wife of Ronald McNair,

their children, Reginald, 3, and Joy, 18 months, and Mrs. McNair's father, Harold Moore; Marvin and Betty Resnik, parents of Judith Resnik; Marsha Jarvis, wife of Gregory Jarvis; Steven McAuliffe, husband of Christa McAuliffe, and their children, Scott, 9, and Caroline, 6.

When it became apparent that something terrible was happening, one of the Smith children was heard to cry out: ''Daddy! Daddy! I want you, Daddy! You promised nothing would happen!''

Seated in the VIP grandstand near launch control, Christa McAuliffe's parents, Edward G. and Grace Corrigan, and two sisters, Lisa and Becky, saw her disappear in an instant that would be an eternity. NASA protocol assistants gathered all the family members from the rooftop and the VIP bleachers and transported them by bus back to the crew quarters. Linda Reppert, Judith Resnik's older sister, joined their parents there and they embraced in tears.

The families were given rooms at crew quarters and served coffee and doughnuts. They rested during the afternoon. Later, Christa McAuliffe's husband, Steve, issued a statement through his Concord, New Hampshire law office: ''We have all lost Christa,'' it said.

In the early evening, Vice President George Bush arrived from Washington with Senators John Glenn (D-Ohio) and Jake Garn (R-Utah) to speak with the families in the crew lounge. June Scobee, wife of the flight commander, spoke for the families. She thanked the visitors for their personal as well as official condolences.

Senator Garn, as was mentioned earlier, had been the first political figure to fly in space. An experienced pilot, he had accumulated more hours in aircraft than had most of the astronauts—probably, he quipped, because aircraft flew more slowly when he flew them. Senator Glenn, the first American to fly in orbit as the United States entered the space age, seemed to be able to empathize with the families more closely than anyone outside their circle of grief.

John Glenn expressed the philosophical outlook of the astronaut corps that enabled them to accept danger and the prospect of death as test pilots. During flight the astronauts were always aware that they were pioneering a new environment, one singularly unforgiving of error or oversight. Glenn told the families that he had felt a sense of loss, akin to theirs, when Virgil I. (Gus) Grissom, Edward H. White II, and Roger B. Chaffee perished in the Apollo 1 fire just 19 years before.[11] The astronauts were an extended family in those days, but even as the corps became larger, a sense of family identification remained.

Glenn himself had been the subject of deep concern for a brief but agonizing period on February 20, 1962. Mercury control feared that he might be doomed to perish on the first American orbital flight. An erroneous signal received at Cape Canaveral indicated that the heat shield had separated from the Mercury capsule, Freedom 7, on the second revolution of the three-orbit flight. If this had occurred, the capsule was in danger of being incinerated by the frictional heat of reentry as it plunged into the atmosphere. Although the signal seen at Mercury Control was not repeated on the spacecraft cockpit display, Glenn was instructed to keep his retrorocket package in place through reentry after the rockets had fired to accomplish the de-orbit maneuver. The package, strapped to the vehicle behind the heat shield, would thus hold the heat shield in place through reentry.

Glenn kept the retrorocket package on, although he did not appear to share the concern of the controllers on the ground. He reported that it made an impressive fireball as Freedom 7 reentered the atmosphere and landed unscathed.

Another moment of stress had occurred on the second Mercury orbital flight when Malcom Scott Carpenter, flying three orbits in Mercury-Atlas 7, overshot his Atlantic Ocean landing zone and disappeared from the space coast radar screens and Mercury Control

radio. After 45 minutes Mercury Control said that he was sighted by a search aircraft that had homed in on his Sarah Beacon, an automatic beeper. He was seated in a rubber life raft that he had hitched to the floating capsule.

The crew of Apollo 13 narrowly escaped disaster when the oxygen storage tank blew out in the Apollo service module during the flight to the Moon in April 1970. Aborting the planned lunar landing, the crew executed a circumlunar maneuver and, with the lunar landing module still attached, flew back to Earth, using the lunar module's oxygen supply and electric power for life support. The crew then made a successful reentry and landing in the Apollo command module.

The beginnings of space flight, like the beginnings of aviation, had not been free of fatal accidents. The Russians had lost four cosmonauts in the Soyuz spacecraft program, one in a crash landing and three as the result of accidental decompression at reentry. Since the shuttle started flying in 1981, its apparent freedom from life-threatening failures had led to the public impression that it was as safe and dependable as a streetcar. The shattering of this illusion amplified the human tragedy that sent a shock wave through the nation.

From the Soviet Union, First Secretary Mikhail Gorbachev sent a message: "We share the feelings of sorrow in connection with the tragic death of the crew of the space shuttle *Challenger*. We express our condolences to their families and to the people of the United States."

Organs of the Soviet press were less diplomatic. TASS said the disaster was a symbolic warning against the militarization of space. *Komsomolskaya Pravda* shook a minatory finger, accusing the United States of setting the rate of flights unjustifiably high to accelerate the strategic defense initiative.

Historically, manned space flight had been motivated by the conviction that it demonstrated technological prowess as well as by a belief that it was the destiny of humankind to expand into the solar system and exploit its vast resources. America had been first to put men on the Moon in this context. Project Apollo had

---

[11] The fire in the Apollo 1 spacecraft was started by an electrical short circuit and ignited supposedly nonflammable plastics while the crew inside was running a prelaunch test on the launch pad. Unable to escape, all three crewmen were suffocated by gases from the burning plastics.

amply demonstrated American technological superiority in the late 1960s and early 1970s; the shuttle was expected to continue it as well as to provide a cost-effective means of transportation to low Earth orbit. In terms of these values, the *Challenger* accident was a national calamity. It was visible worldwide in horrifying detail, demonstrating the innate fallibility of an agency renowned for technical excellence and the fragility of its grandiose plans to extend human enterprise into the solar system.

The day of January 28, 1986, which had dawned on a scene of confident expectation for another successful mission, the twenty-fifth in a row, ended in death and despair. A Florida journalist summed it up: "The relatives now are left to cope with the sight that has linked strangers in grief: Seven people at the pinnacle of their lives, riding a symbol of national achievement in a disintegrating fireball."[12]

By the end of the day, it appeared to me, and others, that the American space program was in a state of collapse.

Although the astronauts who flew the shuttle had maintained that any machine of its complexity was bound to fail sometime, the actuality of catastrophic failure after 24 seemingly routine flights seemed to exceed statistical probability. If the shuttle had been flying with a problem dangerous enough to destroy it, close observers and the astronauts themselves were unaware of it.

However, statistical probability calculations by researchers at the RAND Corporation warned that at least one of the four orbiters in the fleet might be lost to an accident, and that the loss would result in extended grounding of the entire fleet and curtail military as well as commercial and scientific missions.[13] The RAND study had noted 52 anomalies in *Columbia*'s first flight. The study sought to show that the shuttle would have to complete at least 80 missions before it could achieve

the reliability of the expendable rockets that it was initially designed to replace as a satellite carrier.

Response

During the afternoon of January 28, the news center at the Kennedy Space Center press site was in turmoil. Correspondents swarmed at the long counter demanding information, riffling through miles of outdated handouts, and exchanging speculation. Newspapers without correspondents at the scene kept the telephone lines busy, asking for eyewitness accounts, explanations, and accident details. The lines at the Kennedy and Johnson news centers were jammed. The public affairs officers were helpless; they knew no more than the correspondents. NASA officials were either grimly silent or not available.

To the clamor by news people for some official explanation of the accident and its consequences, NASA replied by calling an afternoon news conference at the Kennedy Space Center. Other NASA centers—Johnson at Houston, Marshall at Huntsville, Alabama, the Jet Propulsion Laboratory at Pasadena, California, and NASA headquarters, Washington—were linked to Kennedy by satellite and by telephone lines.

At 4:30 P.M., the Kennedy Space Center information chief, Hugh Harris, opened the conference in the press site auditorium and presented to the packed house Jesse Moore, Associate Administrator for Space Flight in the headquarters hierarchy. A few weeks earlier, Moore had been appointed director of the Johnson Space Center, but he retained his headquarters post pending a replacement.

Moore, 46, held a master's degree in engineering from the University of South Carolina. He had been serving as associate administrator since the summer of 1964, having ascended the administrative ladder from the Office of Space Science at headquarters and from the Jet Propulsion Laboratory. He was known to the "regular" correspondents, that is, the men and women

[12] Jay Hamburg, *Orlando Sentinel*, February 9, 1986, p. A-16.
[13] Article by David Leinweber of RAND in *Machine Design*, March 6, 1986.

who usually covered space activities, for his factual, straightforward briefings, which were usually free of rhetoric. Now he faced a tense audience highly charged with emotion, anxiety, frustration.

"It is with deep, heartfelt sorrow that I address you here this afternoon," Moore said. His voice was trembling. "At 11:30 A.M. this morning, the space program experienced a national tragedy with the explosion of the space shuttle *Challenger* approximately a minute and a half after launch from here at the Kennedy Space Center.

"I regret that I have to report that based on very preliminary search of the ocean where *Challenger* impacted this morning—these searches have not revealed any evidence that the crew of *Challenger* survived."

He recited the crew roster: Francis "Dick" Scobee, commander; Michael J. Smith, pilot; Dr. Judy Resnik, Lieutenant Colonel Ellison Onizuka, and Dr. Ronald McNair, mission specialists; and the payload specialists, Christa McAuliffe and Greg Jarvis. After a pause, he cleared his throat and went on:

"All early indications in the launch control center at the Kennedy Space Center had indicated that the launch was normal up to approximately 11:40 A.M. this morning, about a minute or so into the flight. Flight controllers in the launch control center and in the Mission Control Center in Houston were polled immediately after the explosion and reported that they did not see anything unusual up to that point.

"The solid rocket booster recovery ships were immediately dispatched to the area, approximately 18 or so miles down range from Kennedy, along with various Coast Guard and military ships, helicopters, planes.

"I have taken immediate action to form an Interim Investigating Board to implement early activities in this tragedy. Data from all the shuttle instrumentation, photographs, launch pad systems, hardware, ground support systems, and even notes made by any member of the launch team and flight ops [operations] team are being impounded for study.

"A formal board will be established by the acting administrator very, very shortly. Subsequent reports on this tragedy will be made by this formal review board. I am aware of and have seen the media showing footage of the launch today from the NASA Select System.[14] We will not speculate as to the specific cause of the explosion based on that footage. It will take all the data, careful review of that data, before we can draw any conclusions on this national tragedy."

The Interim Mishap Review Board, as it was formally titled, included Richard G. Smith, 56, director of the Kennedy Space Center; Arnold Aldrich, 49, shuttle project manager at the Johnson Space Center; Walt Williams, 63, a NASA consultant and former executive of the Mercury and Gemini programs; William Lucas, 63, director of the Marshall Space Flight Center; James Harrington, 47, director of the shuttle program integration office at NASA headquarters; two astronauts, Robert Overmyer, 49, and Robert L. Crippen, 48, the latter deputy chief of the astronaut office at the Johnson Space Center. Overmyer had flown two shuttle missions and Crippen, four. Representatives of the National Transportation Safety Board were asked to assist.

NASA Lost Its Head

Jesse Moore's attempt to explain the *Challenger* catastrophe and what NASA would do about it at the first postaccident news conference quickly became an exercise in rhetoric. At the time, neither Moore nor anyone else knew the cause of the fireball and explosion. What NASA would do was another imponderable for the space agency had lost its head. James M. Beggs, 60, NASA's energetic and resourceful administrator, was no longer in charge. He had taken an unpaid leave of absence in December to defend himself against a federal grand jury indictment on charges growing out of his former executive role at the General Dynamics Corporation. The indictment charged that General Dynamics,

[14] A voice and video broadcast system showing portions of a shuttle flight.

Beggs, and three other company executives had tried to cover up cost overruns on a fixed-price contract awarded by the Army in 1978 to build two prototypes of the so-called Sergeant York anti-aircraft gun.[15]

These charges had no connection with Beggs' stewardship of NASA, where he had been serving as administrator since 1981. The Justice Department dropped all charges against Beggs, three codefendants, and the company, and the indictment was dismissed in federal court on June 19, 1987. Government prosecutors admitted that their legal theories and interpretation of records on which the indictment was based were flawed. Dynamic and innovative, Beggs had won President Reagan's approval of the development of the permanent manned space station as the nation's next major space objective. When Beggs went on leave December 2, 1985, his newly appointed deputy, William R. Graham, 49, became acting administrator. Graham, with a Ph.D. in electrical engineering, had a think-tank background but no administrative experience in NASA. After six years at the RAND Corporation, he formed his own research and development firm and served as consultant to the Office of the Secretary of Defense.

Graham also had been an adviser on scientific and technical matters in President Reagan's 1980 election campaign. He was nominated by the President for the NASA deputy administratorship on September 12, 1985 and confirmed by the Senate the following November. He took office on November 25 and a week later, as Beggs went on leave, Graham became acting administrator—to face the worst crisis in NASA's history. Traumatized by the accident, the agency wallowed in rough seas like a ship that had lost its rudder.

If news media folk expected a quick analysis of the accident and a firm prognosis from Moore under these conditions, they were left dangling on the threads of their own speculations. These ranged all the way from sabotage to a blowup of the orbiter's main engines.

---

[15] Ford Aerospace Communications Corporation ultimately won the anti-aircraft gun production contract, but accuracy problems prompted its cancellation on August 17, 1985.

Those engines were suspect because they had exploded with disheartening frequency while being tested in the late 1970s.

Customarily, postflight briefings are followed by questions from the correspondents. Harris said: "Mr. Moore has time for just a couple of questions from each [NASA] center before returning to the effort in investigating this tragedy."

Television correspondent, Jacqueline Boulden, channel 6, Orlando (CBS), asked Moore about reports that *Challenger*'s ascent had appeared to be slower than previous ascents and that there had been an unusual, loud noise that faded and then resounded.

"I've not heard any reports relative to that effect that you described," said Moore. "None whatsoever."

The 51-L mission was the first to be launched from pad B at Launch Complex 39, which had not been used since the Apollo era. The 24 previous shuttle flights had been launched from pad A, about two miles away. Thus, correspondents were viewing the initial ascent of *Challenger* from a new perspective. Still, other observers said they had the impression that the ascent was less energetic than usual. Was this an illusion? Moore thought so. Flight data received at the Johnson Space Center's Mission Control said the ascent was nominal (normal).

Albert Sehlstedt of the *Baltimore Sun*, who had been covering space programs since Project Mercury, asked Moore for an estimate of the volume of liquid hydrogen remaining in the tank at the time of the explosion and what its explosive power was in terms of TNT.

Moore didn't know. "You realize what we've been doing since 11:40 this morning is, we immediately pulled out senior management together on this program and I formed an interim board to ensure all the relevant data to this event would be impounded and would be made accessible to the investigative people that will go and take a look at it. I can't answer your specific questions relative to how much fuel is on board at this point in time. When the board, when it is formally formed by the acting administrator, I'm sure [it] will go into those kinds of questions, but I can't answer it right now."

William Hines, the chief of the *Chicago Sun-Times* Washington Bureau, who also had covered space since Mercury, came on the line from the Jet Propulsion Laboratory in Pasadena, where he and another group of writers had been observing the Voyager 2 encounter with the planet Uranus.

Referring to reports Moore mentioned from the interim review board, Hines said: "Not to put any adverse interpretation on this, it sounds a little bit like a news blackout, and I'm wondering if your objective is to consolidate information. What will be the point of issue of all the announcements in the future? And are other people, not involved in that review board, forbidden to talk?"

"Well," said Moore, "let me correct your statement. First of all, I said that I had appointed an interim review board. That interim review board is composed of senior members of the NASA team here to take immediate actions on impounding data. The acting administrator is expected to appoint a formal review board very shortly, and it will be left up to that review board to determine its progress reports in terms of their findings."

When the press questioning shifted by satellite to reporters assembled at NASA headquarters, Washington, Moore was asked if there was any indication of sabotage or attack on the shuttle. The question was not answered. "I'm sorry, we didn't read that question," said Harris at Kennedy. "And while they're repairing that circuit in Washington, we'll go to the Goddard Space Flight Center in Maryland." The sabotage question was not repeated.

Moore was asked about the weather at the launch site. This was to become a central issue in the investigation. "Were there any unusual weather conditions aloft, or any unusual weather conditions during the launch?"

"None that I recall," said Moore. "We did put up some weather balloons this morning. We did look at load conditions as we normally do, and winds aloft looked good. We didn't have any excedences as far as load indicators are concerned to my knowledge and we thought everything was in good shape for a launch this morning."

The fact that the weather on the morning of January 28 was the coldest in which a manned spacecraft had ever been launched from Florida was not mentioned.

When the around-the-country question-and-answer circuit was completed, the interrogation was resumed briefly at Kennedy. Another veteran newsman from the Mercury days, Jay Barbree of NBC radio, came to the point: "Jesse, I'd like to know if you know what happened to *Challenger*? Can we assume that it was consumed in the explosion? And can you tell us specifically what you know now that the recovery crew has recovered in the impact area?"

Moore replied: "I have not gotten a briefing, Jay, on what the recovery team has found at this point in time. And I have basically looked at the NASA Select photos and so forth, as you did, and all I can say is that it appeared from those photos that there was an explosion. And that's about all I can say at this point in time."

In response to several questions about possible pressure put on NASA to launch and clear the pad for the next mission, Moore said: "There was absolutely no pressure to get this particular launch off. All of the people involved in the program to my knowledge felt that *Challenger* was quite ready to go. And I made the decision along with the recommendations from the teams supporting me that we launch."

By the end of the day, two questions had become paramount. What caused the accident? What had happened in NASA to generate the cause?

In the view of veteran space reporters, the process of disseminating information was a crucial point in the investigation, especially when the image of the United States as the world's leading space power was at stake. Despite the requirement of full disclosure of NASA activities by the National Aeronautics and Space Act of 1958, the space agency held as tight a rein on information it considered negative as did any other federal agency. In 1967, the cause and effects of the Apollo fire and the actual cause of the deaths of crew were withheld for months until the official NASA investigation was completed and reviewed. NASA officials indicated that they would follow that policy in the *Challenger* accident. Inasmuch as news media people had no inkling of the cause of the accident, the news blackout was total and, at the time, seemed likely to last for months.

Meanwhile, NASA and contractor engineers pored over launch films that had not been shown to the public, from automated cameras along the coast. If there were any clues, they might be found there.

2. T Plus 24 Hours

President Reagan reacted to the *Challenger* accident by preparing to deal with it as a national disaster. In that context, management of the investigation was to shift from the leaderless National Aeronautics and Space Administration to a blue-ribbon commission appointed by the White House.

Meanwhile, the President's first declaration was to assure the nation that the American space program would recover and go on as, he said, the *Challenger* crew would have wished. By midday January 29, the broad-scale consequences of the accident were clear. The Space Shuttle Transportation System was grounded indefinitely. The 1986 flight schedule, including two launches to Jupiter and the launch of the Hubble Space Telescope, was put on hold.[1] The space telescope launch has been tentative; now it was indefinite. Commercial and military satellite cargoes on the 1986 manifest might be shifted to expendable rockets—Delta, Atlas Centaur, Titan III.

From its near-equatorial launch site in French Guiana, the European Space Agency's launcher, Ariane, would now acquire a larger share of the market for launching communications satellites. Launched by France, a principal developer, it was a successful competitor with the shuttle as a satellite carrier. Its commercial management, Arianespace, offered lower prices.

At noon on the 29th, NASA presented Acting Administrator Graham to the press at the Kennedy Space Center. With him on the dais were the center's director, Dick Smith; Associate Administrator Moore; and Lieutenant Commander James Simpson of the Coast Guard. Graham announced that President and Nancy Reagan would attend a memorial service for the *Challenger* crew at the Johnson Space Center, Houston on January 31. It was the only hard news Graham had to offer.

"While we grieve for our lost colleagues and for the families that they have left behind," he said, "the NASA team is dedicated to understanding very thoroughly what occurred yesterday and what is needed to give all of us the best possible assurance that it will never happen again. We are dedicated to pressing on."

About 200 news correspondents listened to the statement in silence. There was nothing else to report. The correspondents then began to probe for answers that would relieve the tension.

Graham was asked by the *Atlanta Journal-Constitution* correspondent, Lee Hucks, what the loss of a quarter of the shuttle fleet and the indefinite flight postponement would mean in terms of the nation's security commitments.

"Lee, we have not had the opportunity to go forward with our plans far enough to reach any long-range conclusions at this 24-hour point after the accident," Graham said. "Right now our entire effort is focused on acquiring all the information . . . in times to come we will certainly have to address that question, but not yet."

Again, Graham was asked about the cold weather.

---

[1] Jupiter was the goal of Galileo, an instrumented spacecraft that was to fall into orbit around Jupiter and drop a probe into the atmosphere. Ulysses, a solar observatory, was to be accelerated by Jovian gravitation into a solar orbit passing the poles of the sun.

Reports were circulating that Rockwell International, the orbiter's prime contractor, had urged a delay in the launch because of ice conditions on the launch pad. Rockwell's concern, it was understood, was that icicles hanging from the pad structure would be shaken loose by the launch and damage the brittle silicate-based tiles that form the orbiter's heat shield, exposing the vehicle's aluminum skin to the high heat of reentry and also to aerodynamic heating during ascent.

The question of whether Rockwell had suggested a launch delay, later to become critical, was left in limbo. Graham referred it to Moore and Smith.

Moore said: "There were a series of technical meetings yesterday morning about the ice on the launch pad. The ice team went out and did an inspection early in the morning. And then came back and reported. And the technical people did sit down—all the NASA people involved as well as the contract people involved—and did feel that the conditions at the launch pad were acceptable for launch and basically recommended that, you know, we launch.

"In addition to that, there was a follow-up activity very close to launch. We sent an ice team out about 20 minutes before launch and basically checked again about the ice conditions. And all our reports back from that team were good."

Moore asked if Smith had anything to add. "No, I think that is the case," Smith said. "It was very thoroughly assessed and the indications were that there was very low risk involved."

Had Rockwell requested a launch delay? Press reports said the prime contractor had done so. Such a request implied that the contractor believed it was not safe to launch *Challenger* while ice threatened the integrity of the heat shield. Had such a request been ignored? Misunderstood? Overruled? In matters of launch safety, a no-go recommendation by a prime contractor was tantamount to law. It appeared that the report of Rockwell's concern was not an issue the board was prepared to discuss.

What could be discussed? "Gentlemen," asked Chet Lunner of Gannett News Service and *Florida TO-DAY* newspaper, "could you tell us if, since the accident, what type, what size, what nature of evidence we've collected so far? Whether you've got any leading theories, and when will the board be reporting to us and by what method?"

"Well," said Moore, "let me tell you what I have done with my Interim Board." He named the members appointed on the 28th and said: "We have been meeting almost continuously since yesterday morning and forming a number of different teams . . . to make sure that all aspects of the flight from a data gathering point of view, from an impoundment point of view, are protected."

For example, he said, there was a "set up" to examine flight telemetry; a main propulsion system team; a launch pad–beach area team to search for debris; a team to look at all the photography and television imaging of the launch; and a team to look at all the hardware processing done on all elements of the flight and particularly on the orbiter, the solid rocket boosters, the (external) tank, and the engines from a pedigree standpoint.

"All that is in place," he said. "I've also set up a flight crew team, and the astronaut corps has come forward in a very strong, positive manner and we've got them integrated into our team across the board right now."

Inasmuch as there did not appear to be the slightest clue to the cause of the disaster, Moore had laid the foundation for a long investigation that would put the shuttle transportation system on hold for at least a year.

Harry Kolcum of *Aviation Week and Space Technology* asked whether telemetry data had given "any indication so far . . . that would change what we saw visually. Was there something behind the plume that the telemetry got that we didn't see?" The functioning of all major flight systems and the response of the orbiter to environmental conditions, temperature, and pressure during the ascent were routinely radioed to Mission Control, Houston. This telemetry system stopped at 73 seconds. Controllers perceived no indications of any anomaly in the telemetered data displayed on their consoles. Would more detailed analysis shed any light on what had gone wrong?

"Harry, as I reported," replied Moore, "we have not had a report from any of our teams. They are off making sure that we preserve the data and getting copies of the telemetry data so the right people can look at it from an analytical standpoint. We do not have reports back. . . ."

The correspondents kept probing. "This is for Mr. Moore," said Bill Braden of the *New York Times*. "This question goes to how good the flight was up to the point of the explosion."

Braden said when he compared the ascent timeline prepared in advance with calls during the actual ascent, he noticed that *Challenger* hadn't ascended as far as the prepared timeline calls had indicated. That suggested the ascent had been slow.

"This is not the case from what I have heard to date," said Moore. "All I can say is the reports that I hear on the net, and from the people I've talked to, the flight looked essentially nominal during the planned roll program. During the throttle-down where we throttle down the main engines to about 66 percent . . . and then we all heard the call of throttle back up to 104 percent . . . that went nominal as far as I could tell."

Braden: Were there any unusual anomalies whatsoever, significant anomalies in that period up to the point of explosion that you know of?

Moore: Not that I have heard, quite frankly . . . I have not heard of any significant anomalies during that period of time. Like I said, my inputs as of now indicate that the flight was normal up to that point in time. Now, I have not looked at the data in detail; we have not gone back and run the analysis on all the flight trajectory data in great detail. That's what we're trying to set up and make sure we've got a mechanism to do that . . . and just as soon as we understand that and we can talk on a conclusive basis as opposed to a speculative [one], I think we'll be able to come forward.

Recovery at Sea

Lieutenant Commander Simpson of the Coast Guard was the only official at the briefing able to provide factual information. He said that Coast Guard vessels had begun searching the ocean offshore Cape Canaveral to about 20 miles for debris during the afternoon of the 28th. Having fallen from altitudes above 9 miles, debris from the explosion was widespread, and much of it was floating.[2]

By midmorning of January 29, Simpson said, there were 5 Coast Guard and U.S. Navy ships and 9 Coast Guard, Navy, and Air Force aircraft searching the impact area over a radius of 40 miles from Cape Canaveral. Simpson said that he was anxious to pick up every scrap of shuttle debris before it sank. When the surface had been cleared, the underwater search would begin.

Coast Guard vessels and NASA's booster recovery ships had begun picking up debris by sunset on the 28th and continued to do so at night with powerful search lights. The material picked up on January 28 and 29 amounted to several tons. It was brought into Port Canaveral for inspection and identification and then hauled by truck to a hangar on the Cape. Later, an attempt would be made to reassemble some of the pieces into a meaningful pattern that would show how the shuttle broke up. This would be done by NASA and contractor engineers with the assistance of experts from the National Transportation Safety Board.

So began the largest single salvage operation in world maritime history. By February 5, 28 Coast Guard, Navy, and NASA ships had picked up more than 12 tons of debris in an ocean area of 6,300 square nautical miles. The lot included the orbiter's rudder and attached speed brakes, a big section of one cargo bay door, parts of the aft and forward fuselage and of the main engine bulkhead. The main engines, the solid rocket booster segments, and pieces of the external tank were on the ocean floor by then, below the moving current of the

[2] Radar data examined later showed that *Challenger* broke up at 104,000 feet (19.7 miles) and that some debris reached a peak altitude of 122,400 feet (23.19 miles).

Gulf Stream. The search for floating debris extended from the Cape to Charleston, South Carolina and was carried on from the skies by 13 helicopters and 4 fixed-wing aircraft.

During the morning news conference on the 29th, in response to questions from Roy Neal of NBC and Michael Mecham of Gannett News Service, Graham tried to project the shape of the investigation. He made it clear that the Interim Mishap Review Board named by Jesse Moore was only a start.

"We will undoubtedly as time goes on, at a national level, establish further means to explore and investigate this," he said. "I doubt that it will be focused entirely on one investigative group. We're going to try to draw on a broad range of talent, on consultants, on advisers, on experts, on people from the academic, the industrial, the flight world. We'll put all that together in a number of structures and provide the best national capability to study this, to analyze it, to find out how to correct it, and to ensure that it will never happen again."

"Bill, where will you pull it all together?" Neal asked.

". . . in Washington," Graham replied. "It will be pulled together at the national level throughout the administration. NASA will play a major role in the process."

This indication that NASA would play a major role but not run the investigation suggested that NASA itself would be investigated and its management of the national space program examined closely. It could be surmised from Graham's cautious remarks that the President and his advisers would determine the form of the inquest, leaving NASA the role of gathering data.

But before speculation on this point went further, Graham was asked whether the crew had been able to perceive any indication of an impending explosion. Could they have seen a flash? asked Pete King of the *Los Angeles Times*.

"There was no indication that the commander would have some sense of it," Graham said. "Nor that the ground crews could have had some sense of it in the launch control facilities or the flight operations. You have to understand that this shuttle was traveling on the order of 1,800 feet a second, something over mach 2, twice the speed of sound. It was at 47,000 feet, essentially at the boundary of the stratosphere. A great number of phenomena occur there which involve physical effects we don't see in everyday life. I can assure you you'd see phenomena on the outside of the shuttle on every flight."

During the late afternoon of the 29th, a second news conference was held, this time at the Johnson Space Center. Jay Greene, the flight director at Mission Control during the *Challenger* ascent, and Steve Nesbitt, the commentator who had described the ascent and announced the explosion, told what they knew.

"The prelaunch countdown as far as the Mission Control Center was concerned was perfectly nominal," Greene related. "We had no malfunctions of any significance that we were dealing with. There were no requirements waived either in the systems area or in the weather area. It was probably one of the better weather days we've ever experienced in flight.

"First, a couple of words about the way we operate. We operate by intervening as seldom as possible with the crew. We intervene for probably three reasons: to assure communication, to relay system status back and forth as required, and to pass any data such as abort boundaries that get later in the flight, that are derived on the ground and not available on board.[3] Outside of those calls, any calls we make—we don't say anything unless there is something to say.

". . . calls that we have during the first-stage flight are a confirmation that the roll program has begun and is looking good, which is as much as a com [communications] check as any other check. A call when the throttle sequence is complete and that call which I know has been misquoted is 'go at throttle up' and indicates that all systems on the vehicle including the engines are

---

[3] Abort boundaries refer to emergency landing sites the orbiter can reach under certain conditions of main engine malfunctions and ascent velocity.

looking good to us in the MCC [Mission Control Center]. It is not a 'go' for the commander to throttle up. The event was completed satisfactorily.[4]

"Later in the first-stage flight there are two other calls. One is PC-50, which indicates that the SRBs [solid rocket boosters] are nearing the end of their scheduled burns and is used for some ground monitoring on the second stage [orbiter main engine]. And the next call is the report from the crew that solid rocket boosters have finished their separation.

"The two calls we got through yesterday were go at throttle up. All the calls were nominal both going up and coming down. All of the data we looked at in real time—and we played it back a couple of times afterward—looked nominal. Every measurement that came into Building 30 [Mission Control] was in the range of normalcy as far as any of our operators were able to tell in real time. There were no anomaly reports.

"The way we got the problem in the control center—we had a sharp cut-off of all data. We had the video which you all saw. We had radar tracking which showed multiple pieces and the probability of an explosion. The only thing we were able to do at that point was to coordinate and actually to get ourselves visibility with the search and rescue forces operating out of the Florida area."

Greene said that the search and rescue teams are in readiness at every launch. With the indication of an accident, they were brought into the primary net so that Mission Control could have direct contact with these operations. The Miami Coast Guard took the operational lead, he said.

The mood in Mission Control was "extremely professional under the circumstance," he added. "It was very somber and there was not much said. Everyone watched the TV like everybody else, sort of hoping that maybe something better would come out of it. After a while, we resumed gathering the data, put it all up and called it a day."

[4] The call meant continue at maximum acceleration.

The questioning was opened by Craig Covault of *Aviation Week and Space Technology*. He asked Greene at what point during the ascent he polled the room (that is, called for comment from each controller).

"I never did poll the room," said Greene. "We usually don't poll the room until somewhere around three minutes into the flight, and that's more ritualistic than anything. We have a study loop with all the guys in the room, and if anything happens they come to us as quick as possible. There was nothing to report."

Covault asked Greene to compare the amount of telemetry seen on the control room consoles with more detailed data, recorded in the tapes but not presented on the control room consoles. The more detailed transmissions would have to be analyzed later. There was a general feeling among the correspondents that clues to the cause of the accident would be found somewhere in the tapes.

"The data we looked at was both digital and analogue," Greene said. "The sample rates vary from very high sample rates, like a fraction of a second update rates, up to one-second data and two-second data. Again, none of that data showed anything wrong. I am not familiar with all the data rates that are available to the system and I'm sure somebody, when all that data starts getting pulled together, will give you a breakdown on it."

Bill Hines raised the question of sabotage. "If there had somehow or other been an extraneous signal, either inadvertent or maliciously contrived, that might have led to this event, would that have shown up on your telemetry in any way? And was there any suspicious signal along the line?"

"The only thing I'll say," said Greene, "is we saw nothing anomalous."

"But, excuse me, that's half my question," said Hines. "The first half of the question was, if there had been one would it have shown up?"

Greene: I don't know. What do you mean by an anomalous signal?

Hines: Well, for example, let's just say that some sick, malicious person tried to blow this thing up by sending a signal up to the spacecraft. I know it sound awful.[5]

Greene: We have no evidence that anything abnormal took place with the vehicle.

Tom O'Toole of the *Washington Post,* a veteran space writer, asked why the Air Force range safety officer destroyed both solid rocket boosters when only one was reported to be heading for a populated area. "Do they destroy both when that happens?" Greene said that was the case.

O'Toole then asked whether an explosion of hydrogen (from the external tank) would have a different heat and visual signature than an explosion of solid fuel (from the boosters). Greene said he did not know.

The source of the explosion continued to be a matter of speculation. NASA provided no answers. Again, a question about the weather surfaced. James Fisher of the *Orlando Sentinel* asked Greene: "You said earlier that the weather was good, it was acceptable, but it was pretty cold. When does the cold play a factor in launching the shuttle and what specific concerns do you have about the cold? I understand that there is some concern about ice forming on the [external] tank that could possibly nick the tiles. What are the other concerns, and what specifically do you look for to make sure there will not be a problem in the cold?"

This series of questions was crucial. Cold weather was known to affect the performance of solid fuel rockets, and later events would show that the shuttle boosters were vulnerable to it, but that issue was beyond the purview of Mission Control.

"I think that's a question that can be probably best answered from the Cape," Greene said. "When I was talking about weather, I was talking about things that affect our part of the ascent go—no-go calls. And those things involve the acceptability of the RTLS [return to launch site] and TAL [transatlantic landing] aborts, just in case those were abort modes we would have to rely on."

Albert Sehlstedt of the *Baltimore Sun* reiterated the question of why both solid rocket boosters were blown up by range safety. "Were both headed for the coastline?"

"No," said Greene. "The range safety system is what we call cross-strapped. The range safety people are in the position of guessing what could happen. They have to protect for an out-of-control vehicle [which] could be rotating in any attitude or any random set of attitudes. As a result, they have a requirement for spherical coverage. The only way we can give spherical coverage is to tie both systems together, such that if either system gets the destruct signal, both get it."

Earl Ubell, CBS science correspondent, went further into the possibilities of a clue buried in telemetry detail. "What is it that you can hope to see in the detailed material that you didn't see in the control room?"

"Well," said Greene, ". . . anything that was anomalous would be an indication that may lead the experts, to show where the problem is. The data we look at in the control room is limited to that which is operationally significant, that which we can do something about. Obviously, the external tank temperatures would be very nice to have right now. But operationally, we have no use for it. We have a limited number of operators, and you want them to look at a limited amount of data."

Dick Jackson of *Time* magazine asked whether any changes in crew physiology, such as heightened blood pressure, had appeared in Mission Control Center data immediately before telemetry shut down.

"The only indication I have is that last air-ground transmission, voice transmission," said Greene.

The last words Mission Control heard were uttered by Dick Scobee, saying: "Roger, go at throttle up."

---

[5] Explosive "destruct" packages were attached to the solid rocket boosters and external tank. As mentioned earlier, they could be detonated by a radio signal; the boosters had been destroyed in that manner by the range safety officer after they flew off out of control.

The Apollo Precedent

In the days immediately following the accident, NASA's extensive public information apparatus was overwhelmed by media calls for information, and it virtually had to shut down. The system was created in 1958 under the NASA act that compelled the agency to run an open program. Through the years, it had become oriented toward promoting the agency and creating heroic images. It was not designed to deal with a national disaster.

As mentioned earlier, a precedent for handling news and information in the event of a fatal accident had been set in the Apollo fire of 1967. The agency had maintained an official silence until the investigation was completed months later. It then called a press conference in Washington and presented the findings. This policy was resurrected now.

The Interim Mishap Review Board met on the morning of January 30, and the Kennedy Space Center public affairs director, Chuck Hollinshead, was assigned as public affairs adviser to the board. His job was to prepare a report of what the board was doing on an "as needed" basis. The report would be released to the public only on approval of Jesse Moore, the chairman.

Hollinshead was instructed to prepare guidelines for other NASA centers to follow in their press releases, in coordination with Shirley M. Green, recently appointed public affairs director of the space agency.

Green, like Graham, was a new recruit to NASA. She had been deputy and acting press secretary to Vice President George Bush. Her stewardship of the NASA public affairs apparatus, which is one of the largest in civil government, began in an atmosphere of tumultuous speculation and confusion.

It was not surprising, therefore, that news of the investigation was managed by the investigators.[6] Policy

---

[6] In the past, NASA public affairs had been directed by an official with the rank of associate administrator and with some influence on information policy. The vacancy in that post had been created by the resignation of Frank Johnson, public affairs administrator under Beggs.

wavered. At its meeting on January 30, the board authorized the Coast Guard to ferry a number of reporters and photographers to the salvage area, but stipulated that the captain of the ship must avoid significant objects. It was not clear what these were. On second thought, the board cancelled the cruise lest it interfere with inspections by the National Transportation Safety Board. At the outset of the undersea search, a request by this writer to obtain an account of the salvage activities by a private oceanographic research organization working with the Navy was submitted to NASA for clearance. It was never answered.

In the absence of any factual information from NASA on the progress or findings of the investigation, the major news organizations had to rely on NASA sources talking unofficially and on interviews with contractor engineers. Through that process the American public learned the cause of the *Challenger* disaster.

The Promise

On the morning of January 31, a memorial service for the *Challenger* crew was held at the Johnson Space Center. It was conducted on the lawn in front of Building 16, where 2,000 folding chairs had been set up. When they were filled, another 8,000 men, women, and children occupied standing room behind them. The central figure there was President Reagan. He and Nancy were seated down front between Jane Smith, widow of the *Challenger* pilot, and June Scobee, widow of the commander. Families of the rest of the crew were seated around them. The ceremony was simple and deeply moving, perhaps the most elegiac event of the space age. President Reagan spoke the eulogy, which was carried nationwide by television. He said:

We remember Dick Scobee, the commander, who spoke the last words we heard from the space shuttle *Challenger*. We remember Michael Smith who earned enough medals as a combat pilot to cover his chest.

We remember Judith Resnik, known as J.R. to her friends. We remember Ellison Onizuka who, as a child running barefoot through the coffee fields and macadamia groves of Hawaii dreamed of some day traveling to the Moon. We remember Ronald McNair who said he learned perseverance in the cotton fields of South Carolina. We remember Gregory Jarvis—on that ill-dated flight he was carrying with him a flag of his university in Buffalo. We remember Christa McAuliffe who captured the imagination of an entire nation, a teacher not just to her students but to an entire people. Today we promise Dick Scobee and his crew that their dream lives on; that the future they worked so hard to build will become reality.

The Puff of Smoke

Within 72 hours of the accident, enough telemetry and film data had been recovered by investigators to point their investigation toward a defective field joint in the right-hand solid rocket booster. A significant clue was a puff of smoke that Horace Lamberth, a Kennedy Space Center engineer, saw in launch film as coming from the lower-segment field joint of the right-hand solid rocket booster at a point near the external tank attachment.

Engineers familiar with booster technology could surmise the probable result. The smoke signaled a leak of burning propellant exhaust from the joint where two booster segments were mated. It produced a torch of hot gas at 5,600° Fahrenheit that melted a hole through the half-inch steel case of the rocket segment and perforated a section of the external tank where liquid hydrogen was stored.

This scenario was quickly perceived by NASA engineers and first made public by Jay Barbree of NBC on the "NBC Nightly News," 6:30 P.M. EST, January 30. One of Barbree's sources was the ex–deputy director of the Kennedy Space Center, Sam Beddingfield, who had retired in December.

The next day, the Interim Mishap Review Board released video tape showing the smoke plumes but made no statement about it. The board viewed 16-millimeter films of the launch sequence showing that the hot gas burned through the joint between the aft and the aft-center segments of the right-hand booster, confirming the scenario indicated by the puff of smoke.

There was more to the story. On January 29, 24 hours after the accident, George Hardy, deputy director of science and engineering at the Marshall Space Flight Center, told the board that a detailed telemetry review showed a mismatch in thrust between the right-hand and left-hand solid rocket booster. It grew to a deficit of 85,000 pounds of thrust in the right booster at the time of the explosion.

Controllers at the Johnson Space Center had not observed it on their consoles, probably because the mismatch was within specifications for the shuttle system. The indication of lower pressure in the right booster was significant only as a clue to the accident that followed, and as confirmation of the hot gas leak indicated by the puff of smoke.

Detailed analysis of telemetry tapes at the Johnson Center revealed that at 1.25 seconds before data were lost, the right-hand booster's rate gyroscope indicated that the nose of the rocket was swinging toward the external tank, while the aft part of the booster was swinging away. The aft attachment between tank and booster had burned through. The nose of the booster crashed into the tank and ruptured it at the bulkhead between the liquid hydrogen and liquid oxygen containers. The fireball then erupted. There were data in the tapes showing that the thrust vector control in the right booster caused the nozzle to gimbal 2 degrees to compensate for the loss of thrust pressure. The data also showed that steering corrections were started by the orbiter main engines as the shuttle began to veer off course. The initial escape of hydrogen when the external tank was perforated was confirmed by telemetry that showed a drop in liquid hydrogen pressure at 67 seconds.

On February 1, the board released additional film

First evidence of black smoke indicating a leak of hot gas appears on the right-hand solid rocket booster at 0.445 seconds after liftoff. (NASA)

showing a glow of burning hydrogen spreading along the external tank. The video tape showed the orbiter emerging from the fireball. It was occluded for a time by dense smoke and then reappeared. The crew compartment could be seen falling clear, but it disappeared from the film as it began a long plunge into the ocean. This vision raised questions that were never fully answered about the possibility of crew survival if the orbiter had been equipped with a launch escape system.

On Sunday morning, February 2, Graham appeared on the CBS news program "Face the Nation." Under questioning, he admitted that a leak of hot gas had occurred near a seam joining two segments of the booster.

The Commission

The Mishap Board reconvened on February 3 at Kennedy Space Center to hear reports of teams analyzing flight performance, flight hardware, ground support equipment, the launch pad facility, and the salvage operations. The Coast Guard was reducing the numbers of surface vessels and aircraft it had been using to search

At 2.147 seconds black smoke spreads halfway across the booster. (NASA)

for and pick up debris, most of which had either been recovered by that time or had sunk. Undersea salvage began.

During the meeting, Chairman Moore asked all non–board members to leave the room. He then told board members that President Reagan was going to appoint a commission to review procedures for preparing and certifying a shuttle for flight. The chairman of the commission would be William P. Rogers. Moore advised that the Interim Board would be renamed and function under a new charter, but he said that he expected that the membership would remain intact.

The President signed Executive Order 12546 that day creating the Presidential Commission on the Space Shuttle *Challenger* Accident. With this act, the President formally transferred control of the investigation from NASA to the commission. Its writ ran beyond a determination of the cause of the accident. It extended to a probe of NASA's management of the space shuttle program. Inasmuch as the commission quickly took the position that the space agency could not be expected to investigate itself, the agency's role in the investigation was relegated to data collection and analysis—for the commission. A metamorphosis occurred in public information policy. The commission adopted the policy that all information about the accident and about events

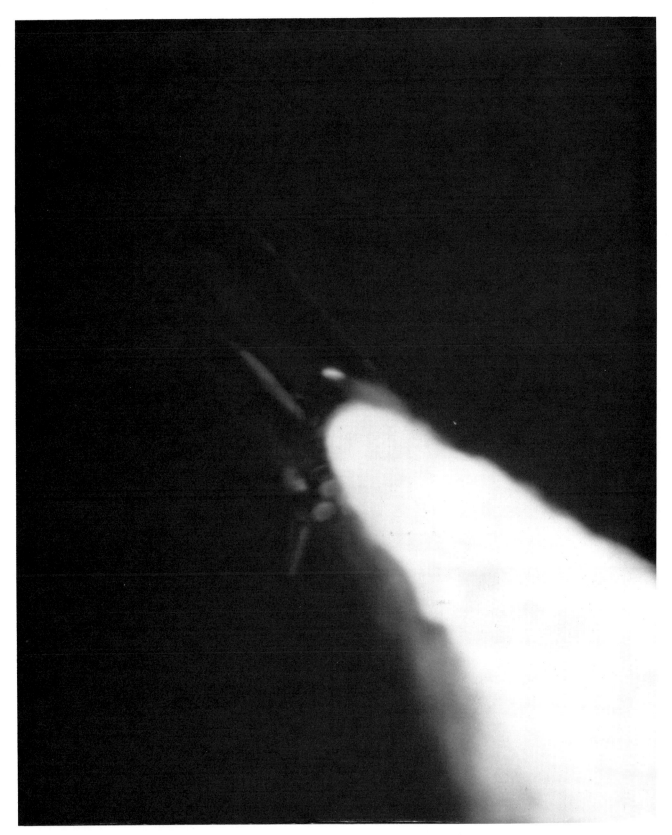

At 59.82 seconds, a still camera photographs an intensifying glow and plume emerging from the lower segment of the right booster. (NASA)

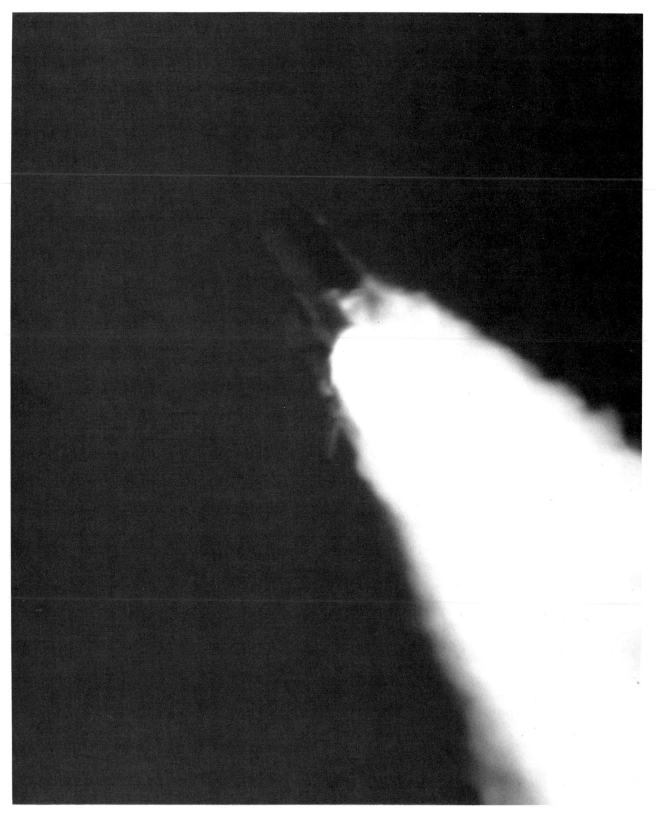

The glow becomes brighter and larger at 66.625 seconds.

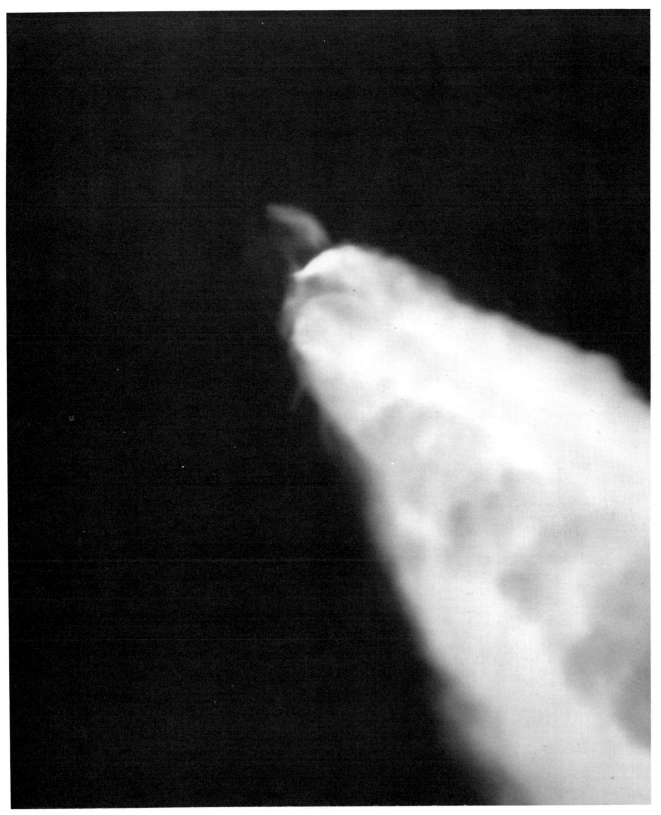

At 73.175 seconds, a bright cloud forms on the side of the external tank. (NASA)

A flash appears between the orbiter and the external tank at 73.201 seconds. (NASA)

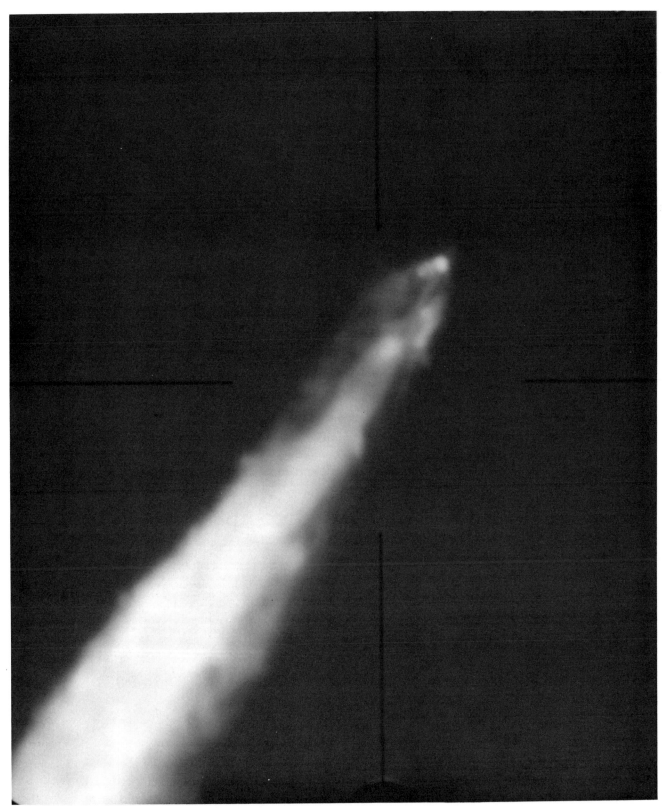

The flash intensifies at the liquid oxygen section of the external tank at 73.326 seconds. (NASA)

Explosion and fireball. The time is approximately 76 seconds in this photo published in the *Commission Report*. At far right, the left booster is seen thrusting away. (NASA)

leading up to it would be made public in a timely fashion through public hearings and the release of NASA documents.

The chairman, William P. Rogers, 72, had served as Secretary of State in the Nixon administration (1969–73) and Attorney General in the Eisenhower administration (1957–61). He was a partner in the law firm of Rogers and Wells, New York. The firm represented the Lockheed Corporation, which held the shuttle processing contract at Kennedy Space Center. Rogers said his

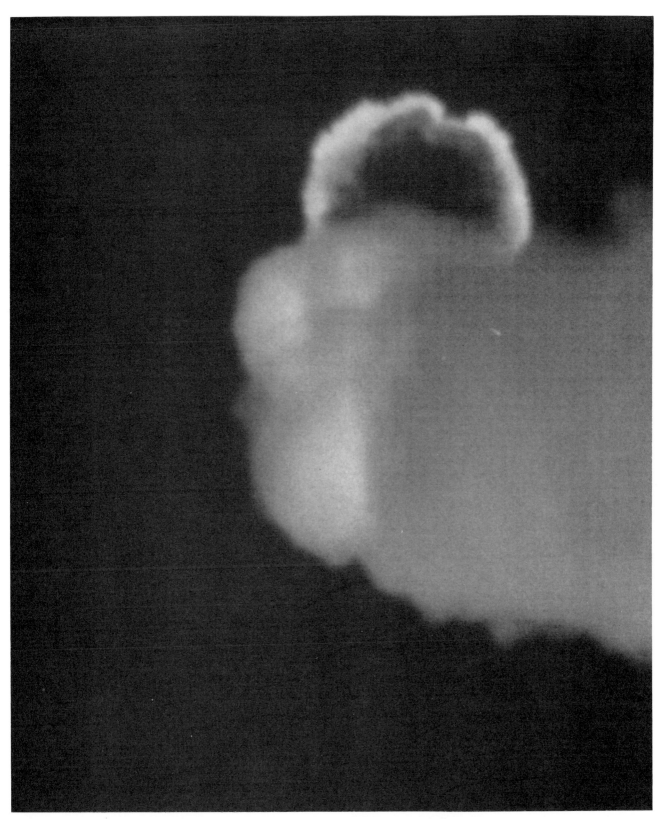

At 109.6 seconds, the right booster is blown up beyond the cloud from the *Challenger* fireball. (NASA)

work at the firm never had anything to do with the space program.

The vice chairman of the commission was Neil A. Armstrong, 55, the former astronaut and the first man on the Moon. He was chairman of the Board of Computing Technologies for Aviation, Inc. of Charlottesville, Virginia. Armstrong had resigned from NASA in 1971 and taught aeronautical engineering at the University of Cincinnati until 1980. He was also a member of the President's National Commission on Space, which had been charged with the task of proposing the nation's future program in space.

Other members, as announced by the White House, were Brigadier General Charles (Chuck) Yeager, 63, Air Force, retired, an experimental test pilot and the first pilot to break the sound barrier, who also was a member of the National Commission on Space; Dr. Sally K. Ride, 34, astronaut and the first American woman to fly in space (aboard *Challenger,* June 18–24, 1983); Dr. Albert D. Wheelon, 57, senior vice president and group president of the Space Communications group, Hughes Aircraft Company; Robert W. Rummel, 70, aerospace engineer and former TWA vice president; Dr. Arthur B. C. Walker, Jr., 49, a physicist at Stanford University and aerospace industry consultant; Dr. Richard P. Feynman, 67, a theoretical physicist at the California Institute of Technology, who received the Nobel prize in physics in 1965 and was a member of the atomic bomb project at Princeton University and the Los Alamos Scientific Laboratory; Dr. Eugene E. Covert, 59, professor of aeronautical engineering at the Massachusetts Institute of Technology, who had chaired a committee of consultants in 1979 to review orbiter main engine problems for NASA at the request of the Senate Committee on Aeronautical and Space Science; Robert B. Hotz, 71, former editor of *Aviation Week and Space Technology,* a leading aerospace journal; David C. Acheson, 64, senior vice president and general counsel of the Communications Satellite Corporation; Major General Donald J. Kutyna, 52, Director, Air Force Systems Command Control and Communications, who managed the design

and construction of the Air Force shuttle launch facility at Vandenberg Air Force Base, California; and Joseph Sutter, 65, executive vice president of the Boeing Company Commercial Airplane Division.

The executive order that spelled out the commission's task stated:

The commission shall review the circumstances surrounding the accident to establish the probable cause or causes of the accident, develop recommendations for corrective or other actions, based on the commission's findings and determinations. The commission shall submit its final report to the President and the Administrator of the National Aeronautics & Space Administration within 120 days of the order of the President.

When the President's press secretary, Larry Speakes, briefed the White House press on the commission, Rogers and Graham were present. Speakes was asked: ". . . well, then, you are saying, I take it, that the NASA investigation now comes to a halt. NASA puts its resources at the disposal of this commission and the commission then proceeds with the investigation. Is that correct?"

Speakes: You've got it.
Q: Well, who provides the technical support then, if not NASA?
A: NASA will provide technical support but this commission will then have the ability to make judgments regarding the information provided them, call others, request additional information from NASA, ask questions of others who are outside of NASA, outside of government, anywhere.
Q: Can we ask Dr. Graham how NASA feels about having this investigation taken away from it? Is it a vote of no confidence, Dr. Graham?
Graham: Well, I don't think it's taken away from them.
Rogers [addressing Sam Donaldson of ABC]: I want to comment on your question, Sam. I don't think it's taken away from them at all. I think we're going to work very closely with them. We're not going to rely solely on that investigation. And because the

President felt that there should be an overall commission that made the final recommendations and the final report, we're going to make it. We're going to make some recommendations; we're going to make the inquiry. Now that is not adversarial as far as NASA is concerned.

On February 4, the 51-L Interim Mishap Review Board met in the early afternoon at Kennedy Space center to discuss its relationship with the commission. At a subsequent "quickie" news conference during a visit to the center, Chairman Rogers insisted that their relationship would be cooperative. At the Kennedy and Marshall Centers, however, there was an atmosphere of foreboding about the role of a commission headed by a former Attorney General. Some people perceived the commission's role as analogous to that of a federal grand jury.

The Cook Memoranda

As the Presidential Commission began its deliberations in Washington, a sensational report that broke through NASA's shell of silence on the causes of the accident appeared in the *New York Times* of February 9. Philip M. Boffey of the Washington bureau disclosed NASA internal documents expressing concern that leakage of hot gas from a solid rocket booster during launch threatened a catastrophic accident.

Boffey cited memoranda from a NASA budget analyst, Richard C. Cook, reporting concern by some agency engineers that synthetic rubber "O-rings" which were emplaced around the circumference of the joints where the booster segments were mated to seal the joints against hot gas leakage had shown increasing incidence of erosion from the gas. These O-rings and putty were installed at the factory and at the Kennedy Space Center to prevent the blowtorch-hot gases from burning through the steel casing at the joints linking the heavily insulated segments. The joints where the segments were mated and the O-rings installed at the

Kennedy Space Center's Vehicle Assembly Building (known as the VAB) were called field joints; others made at the Morton Thiokol factory in Utah where the boosters were manufactured were called factory joints. Because of the boosters' length, 149.5 feet, they could not be shipped to Florida fully assembled. Consequently, they were hauled across country by rail in four segments that were assembled at the VAB by NASA and Morton Thiokol engineers.

Cook had written two memoranda, one six months before and the other shortly after the accident. In his first memorandum, dated July 23, 1985, he advised his superior, Michael B. Mann, chief of the shuttle program analysis branch, that seal erosion was "a potentially major problem affecting flight safety and program costs" in the opinion of engineers he had talked to. It was Cook's judgment that the "budget threat" represented by the O-ring erosion problem was a part of his duties, which included assessment of future program costs.

Although not an engineer, Cook had been amazingly prescient in his warning of catastrophe. Soon after the accident, Cook wrote a second memorandum in which he described the probable cause of the accident. "There is a growing consensus," he said, "that the cause of the Challenger explosion was a burn-through in a solid rocket booster at or near a field joint." That, indeed, was the scenario suggested by launch film and telemetry data.

Cook said further: "It is also the consensus of engineers in the propulsion division, Office of Space Flight, that if such a burn-through occurred, it was probably preventable and that for well over a year, the solid rocket boosters have been flying in an unsafe condition. Even if it cannot be ascertained with absolute certainty that a burn-through precipitated the explosion, it is clear that the O-ring problem must be repaired before the shuttle can fly again."

NASA officials declined to comment publicly on the *New York Times* story. Boffey reported that Moore did not return a telephone call to his home, "although a family member said he was there."

Boffey was quoted as saying that the memoranda were made available to the *Times* by a solid fuel rocket analyst who remained unidentified.[7] There was no indication that Cook was involved or was blowing the whistle. In any case, Cook was leaving NASA to return to a former post in the Treasury Department.

In situations of this kind, where secret or internal documents materialize in a newspaper office, it is customary to explain that some citizen of good repute threw them over the transom in the public interest.

The Rogers commission reproduced the Cook memoranda and made them available to the news media, along with related documents from NASA headquarters and the Marshall Space Flight Center. These papers made it clear that the shuttle disaster was not an anomalous event but was the consequence of a joint sealing system that had shown signs of failing since the STS-2 mission, the second flight of *Columbia*. The miracle was that the shuttle had been launched 24 times before a joint failed.

Since the fall of 1981, evidence that hot gases were eroding the rubber O-ring seals had been seen after the boosters were recovered from the sea for reuse and inspected before being reloaded with propellant at the factory and shipped back to Kennedy Space Center for reassembly. The inspections found that hot gas from burning propellant had blown by the primary O-ring seal and had deposited sooty particles of rubber and grease behind it.

If the primary O-ring were breached, a secondary O-ring was designed to stop the leak before the gas reached the steel casing. But sometimes the secondary O-ring did not work, because pressure of the gas from the burning propellant opened the gap the secondary O-ring was designed to seal so that it would not remain seated in sealing position. If that happened when the primary O-ring failed or burned through, the hot gas could impinge directly on the steel motor case and burn

a hole through it. By the end of the first week of February, it was surmised that this is what happened in *Challenger*'s right-hand booster.

Documents abstracted from the files of NASA by the commission and made part of the public record showed that the O-rings at various booster joints had sustained erosion, blow-by, or signs of heating on half of the 24 shuttle flights before *Challenger* flew for the last time.

The records showed also that the concerns of both NASA and Morton Thiokol engineers were increasing, as the incidence of gas burning or blowing by the seals became more frequent with the acceleration of the flight schedule and erosion increased. A report by Irving Davids, a headquarters engineer, disclosed both primary and secondary O-ring erosion on *Challenger* mission 51-B, launched April 29, 1985. In another instance, to be detailed later, the primary O-ring on a booster failed, but luckily for the crew, the secondary O-ring sealed properly and contained the gas.

The records show that engineers at the Marshall Space Flight Center, the overseer of both the solid rocket booster and the main engine propulsion systems, and development engineers at Morton Thiokol drew up plans to improve seal performance, but did not implement them. The booster sealing design could not be changed substantially without grounding the shuttle, a prospect deemed unwarranted at NASA headquarters. On August 19, a report by Marshall and Morton Thiokol engineers disclosed that erosion had occurred in 18 primary O-ring seals during both flight and static (ground) tests. The report concluded that the primary O-ring in the field joint should not erode through, but if it did leak owing to erosion or lack of sealing, the secondary seal might not seal the motor.

Despite that dangerous prospect, the report recommended that it was safe to continue flying the existing design under specified conditions. A later report, August 30, from Morton Thiokol proposed redesigning the seals to eliminate blow-by and erosion, but nothing came of it.

---

[7] When asked how he got the memo, Boffey maintained the traditional confidentiality about such matters.

Criticality One

The records buried in the files at Marshall and NASA headquarters revealed that the design of the joint seals had been a subject of criticism and concern by some Marshall engineers since 1978, three years before the first orbital test flight. That fact, however, had never come to congressional or public attention. Not even the astronauts knew about it, nor did they know that the seals in the boosters of the shuttles they were flying were a chronic problem.

The solid rocket boosters had invariably been presented by NASA as examples of an old, reliable technology, virtually infallible. A mythology had grown around them. It said they never would fail. The "weak link" in shuttle propulsion was the orbiter main engine system. These hydrogen-oxygen engines were the most powerful and advanced of their type in the world. They had been developed after years of failures and frustration, frequently blowing up during tests. At first, the main engines were suspected as the cause of the Challenger explosion, and Moore's investigators continued to look at that possibility until the evidence pointed directly at the right booster.

The concern of booster engineers about the safety of this huge rocket was not shared widely in NASA. The astronauts, for example, were not aware of it, nor were many officials who were not directly involved in the booster program. But in December 1982, the boosters were designated as "criticality 1" items because the possibility that the secondary O-ring would fail to seal indicated a lack of redundancy. This designation meant that if the primary seal failed during the launch phase, the result would be loss of the mission, loss of the vehicle, and loss of crew "due to metal erosion, burn through and probable case burst resulting in fire and deflagration." The definition of criticality 1 foretold the

fate of Challenger. Shuttles had always flown under that Damoclean sword. In December 1982, the risk was officially recorded in a document that was not made public until the commission fished it out of NASA's files four years later.

Before December 1982, the booster seals had been characterized as criticality 1-R, the R for redundancy. This designation supposed that the secondary O-ring seal would always work. When it was found at Marshall that it would not always work, the safety designation was downgraded to 1.

The reclassification to criticality 1 meant that the shuttle could not fly unless the condition that the secondary O-ring might not seal was waived. The primary design requirement was that shuttle systems should be failsafe: they must be able to sustain a single failure and still fly safely. Failsafe was assured to some extent by redundancy. An example was the orbiter's flight computers through which the spaceship was flown. There were five working in harmony. If one failed, another took over. If four failed, the fifth could fly the ship. If all failed, the ship was doomed.

Some systems could not be made redundant, such as the wing. The booster seals were designed for redundancy but failed to meet the test. Instead of improving them so that they would function redundantly, NASA headquarters issued a waiver of the failsafe requirement. It was argued by some engineers at Marshall that the seals were redundant at least part of the time; there was a high probability that the secondary O-ring would seal if the primary failed.

Crews were thus entrusted to a vehicle system with boosters that might or might not be failsafe and from which there was no escape in the event of "fire and deflagration" during the first two minutes of ascent when the boosters were firing.

3. Anatomy of the Accident

With the establishment of the Presidential Commission, the role of NASA officials in the investigation of the *Challenger* accident was reduced to data collection and analysis for the blue-ribbon panel. Moreover, the manner in which the space agency had developed, tested, and operated the shuttle transportation system became a subject of the investigation. Although the subordination of NASA to an ad hoc commission indicated a lack of confidence in the agency by the White House and the government's Inter-Agency Space Advisory group, Chairman Rogers insisted that this was not the case. In a brief appearance at the Kennedy Space Center, as was noted earlier, Rogers had defined the relationship between the commission and NASA as one of mutual cooperation; but this tactful pronouncement was soon followed by the commission's finding of a flaw in the decision to launch *Challenger* and an order dropping officials involved in it from the investigation team.

The Interim Mishap Board was reorganized by Graham on February 5 as the Data and Design Analysis Task Force, a title that eliminated the euphemism "mishap." The task force continued to function under the chairmanship of Jesse Moore until February 19, when he was succeeded as associate administrator for space flight by Rear Admiral Richard H. Truly, 48, former astronaut and Navy test pilot. Truly had flown the orbiter *Enterprise* on approach and landing tests in 1977; had been pilot on the second orbital flight of *Columbia* (STS-2) on November 12–14, 1981 and commander of the *Challenger* mission (STS-8) that executed the first night launch and night landing, August 30–September 5, 1983. Truly resigned from NASA to head the Naval Space Command, which controls Navy satellites, and continued in that role until he was persuaded to rejoin NASA as a headquarters official.

Moore took up duties in March as director of the Johnson Space Center, Houston, a post to which he had been appointed before the accident to succeed Gerald D. Griffin, who had retired in 1985. Moore had remained in Washington pending the appointment of a successor there. These moves reflected the onset of reorganization of the top levels of NASA administration following the retirement of Beggs, which left agency management in an uncertain if not chaotic state.

One of the first releases of information by the new task force was the official timeline or sequence of events that had finally been authenticated as having occurred on the *Challenger* 51-L mission from main engine ignition to the destruction of the solid rocket boosters by the range safety officer. This was the first step in defining the structure of the disaster.

The main engines started at L (for launch)[1] minus 6.6 seconds, and the solid rocket boosters were ignited at zero (11:38 A.M. Eastern Standard Time). The orbiter engines always start up before the boosters in order to build up to full thrust.

First movement of the shuttle, with a mass of 5.4 million pounds fully fueled, was recorded in the telemetry record at L plus 0.0587 seconds. Then at 0.445

---

[1] Launch time is also designated by T.

seconds appeared the first clue to the accident—a puff of black smoke near a field joint on the right-hand solid rocket booster photographed by camera 60.[2] The photos showed the smoke darkening at 1.6 seconds and extending halfway across the booster at 2.147 seconds. At 7.7 seconds, Scobee began a maneuver to roll *Challenger* to ascent attitude so that the orbiter faced earthward. The maneuver reversed the position of the right booster from right to left relative to observers on the ground, but the smoke continued to be visible for 12 to 13 seconds and then disappeared.

At 20 seconds, Scobee throttled down the main engines to 94 percent of nominal full thrust (100 percent). The roll maneuver was completed at 21.1 seconds. At 36 seconds into the flight, he began to throttle down the main engines to 65 percent of full thrust as the shuttle approached the region of maximum dynamic pressure, "Max Q." This is the region where the velocity of the shuttle and the density of the atmosphere interact to exert peak forces on the vehicle.

At 40 seconds, telemetry showed that *Challenger* was encountering high winds. Then, as Scobee prepared to throttle the engines back up to full power (104 percent of nominal full thrust), smoke appeared at 58.7 seconds. Camera 207 showed the smoke emanating from the side of the right booster just forward of the aft attachment securing the aft end of the booster to the external propellant tank. At 59 seconds when *Challenger* had passed through Max Q the smoke plume became well defined and intense.

At 60.16 seconds, telemetry registered a drop in right booster chamber pressure compared with that in the left booster. A half-second later, flame was filmed by camera 206. At 62.4 seconds, the right elevon (the movable part of the wing) moved and the engines gimbaled, changing the orbiter's pitch, apparently compensating for the drop in right booster thrust. Bright flecks appeared on the bottom and top sides of the booster at 66.17 seconds, and at 66.48 seconds a pres-

sure drop was registered in the liquid hydrogen fuel compartment of the external propellant tank. That was the beginning of the end.

Camera 207 showed a bright, sustained glow on the side of the right booster at 66.62 seconds and the merging of plumes on both sides of the booster. During this time, the orbiter main engines were thrusting, drawing propellant from the tank. At 67.68 seconds, telemetry registered a decrease in liquid oxygen inlet (tank to engine) pressure. At 72.14 to 72.28 seconds, telemetry showed divergent pitch and yaw rates by the right and left boosters and lateral acceleration by the right booster. Hydrogen and oxygen inlet pressure was falling at 72.8 seconds, indicating escape of propellant from the tank. AT 73 seconds, chamber pressure in the right booster fell 24 pounds per square inch lower than that in the left booster, indicating escape of exhaust gas from a hole in the side of the booster.

A white cloud appeared along the side of the external tank at 73.175 seconds, and a fraction of a second later cameras 206 and 207 showed a bright flash from the region between the orbiter and the hydrogen portion of the external tank. This was followed immediately by an explosion near the point where the right booster was attached to the forward end of the tank. At 73.326 seconds camera 206 showed a high-intensity white flash along the forward section of the tank where the liquid oxygen chamber was located. The tank then came apart.

Analysis of all the photographic data yielded this explanation of what happened: A hot gas jet at a temperature of 5,600° Fahrenheit had pierced the joint between the aft-center and aft segments of the right booster and with blowtorch intensity melted or severed the aft attachment between the booster and the tank.[3]

---

[2] Black smoke indicated burning rubber seals and grease in the joint.

[3] The lower attachment ring that girdled the booster near the aft field joint was bolted to the steel case and connected to the external tank by three struts. In a memorandum of December 29, 1986, John W. Young, chief of the astronaut office, said that unexpected high-altitude wind shears played a part in the accident. The catastrophic failure of the booster joint "appears to have resulted from damaged seals

Still firing, the booster rotated on the forward attachment and its nose crashed into the tank, rupturing the liquid hydrogen and liquid oxygen sections. The escaping propellant promptly expanded into a huge fireball.

Although this scenario seemed to be consistent with telemetry and film records, the NASA task force did not confirm it, referring only to events as they appeared in the timeline. The sequence implied that the accident started with a leak in the right booster field joint near the attachment fixture. What remained to be determined was the cause of the leak. That became the focus of the most intensive technical investigation of the space age. After 120 days of examination, the Presidential Commission determined that the leak resulted from the failure of rubber O-rings to seal the joint between the aft and aft-center booster segments. But the precise cause of that failure could only be conjectured.

News media reporters who interviewed Marshall Space Flight Center engineers at Huntsville, Alabama and contractor engineers at the Morton Thiokol manufacturing plant at Brigham City, Utah reported opinion at both centers that the rubber seals lost their resilience in cold weather and might not have seated properly at low temperatures. Although the low-temperature effect was not conceded by NASA officials, it quickly became suspect as a contributing cause of the seal failure. After a night of subfreezing weather, Challenger was launched at an outside temperature just 4 degrees above freezing—the lowest temperature at which any manned space vehicle had ever been launched in America. This circumstance became a central issue in the commission's investigation.

reopening at 58.8 seconds," he said. He attributed this effect to "a weakened external tank attach ring reacting to loads from a significant vertical wind shear." Young suggested that the attach ring might have been weakened when bolts were broken by high loads at liftoff. Broken attach ring bolts had been found in boosters recovered from early launches, he said. However, whether they were broken at liftoff or at ocean impact had not been determined.

The Jupiter Effect

Why was Challenger launched on the coldest morning in NASA launch experience, a day when icicles were hanging from the pad service structure? Was there pressure to get Challenger off the pad? The space agency had scheduled 15 launches in 1986, compared with 9 in 1985, and planned to launch 19 missions in 1987. Delays in the launch of Challenger and, earlier in the month, of Columbia, were threatening to disrupt the schedule.

Rumors of pressure to launch from people on the White House staff were picked up by members of the Senate Subcommittee on Science, Technology, and Space and by members of the Presidential Commission. This was publicly denied by White House spokesman Larry Speakes and by every NASA official who was questioned about it by members of the Senate subcommittee and the commission.

There was no question that launch delays during the unusually cold winter of 1985–86 in Florida were eroding NASA's ability to meet commercial and military satellite launch commitments. News media critics tended to view launch delays caused by engineering problems as evidence of inadequate quality control and caused by weather as indicative of excessive caution and lack of confidence. These views could be attributed to the NASA public information process, which since Project Mercury had sought to create an agency image of infallibility and derring-do.

One source of pressure to launch on the coldest day was not human. It was the planet Jupiter, target of two scientific spacecraft to be launched in the spring of 1986, in order to encounter the big planet on a minimum energy trajectory. Scheduled for launch on May 15 aboard Challenger was the International Solar Polar Observatory, a joint effort by NASA and the European Space Agency to observe the poles of the sun. That region had never been seen by astronomers or solar physicists. The mission had been renamed Ulysses, in reference to a passage in the canto 26 of Dante's *Inferno*, wherein the hero of the Trojan War expresses a desire to explore the uninhabited world behind the sun.

In order to view the poles of the sun, the Ulysses craft had to fly high above the ecliptic plane in which the planets orbit the sun. Such a journey required more energy than could be provided by the Centaur booster after it was deployed by *Challenger* in low Earth orbit.

From there, the flight plan called for the Centaur rocket to boost Ulysses on a flight path passing Jupiter so that the gravitational field of the big planet could deflect the spacecraft into a high-inclination orbit that would carry it over one of the poles of the sun two and one-half years later. Ulysses would then take up a polar orbit of the sun, passing the opposite pole about eight months later. The navigation technique was ambitious but not new. The gravitational field of Venus had been exploited to hurl Mariner 10 into a fly-by of Mercury in 1973. Jovian gravity had boosted Pioneer 10 (1972) out of the solar system, Pioneer 11 (1973) to Saturn, Voyager 1 to Saturn, and Voyager 2 to Saturn, Uranus, and on a heading toward Neptune after their launches in 1977. In 1985, gravitational assist from the Moon accelerated the International Comet Explorer (formerly ISEE-3) out of the Earth-Moon system to an encounter with comet Giacobini-Zinner.

The second flight to Jupiter was Galileo, a mission designed to drop an instrumented probe into the Jovian atmosphere and establish an orbital observatory around Jupiter. Galileo was to be launched May 20 aboard *Atlantis,* which like *Challenger* was also being refitted to carry the hydrogen-fueled Centaur in cargo as Galileo's upper-stage booster.[4]

Ulysses had been scheduled for launch from pad A on the Ides of May, and Galileo was to lift from pad B five days later to fit the Jupiter launch window, which opened at 13-month intervals. For this reason, *Challenger* and *Atlantis* had to be processed simultaneously. It would be the first dual launch in the shuttle experience and a critical test of the flight system.

The timing of these ambitious and exacting missions was being threatened by winter weather delays on the launch pad, first of *Columbia* in late December and early January and then of *Challenger* in January. The Jovian launches posed serious safety problems. Centaur was to be the first liquid fuel rocket to be carried in the shuttle cargo bay. A special loading facility had to be erected at the launch pad for fuel and oxidizer. Both Ulysses and Galileo carried radioisotope thermal generators to supply electrical power over their multi-year journeys. The generators would be fueled by 46.2 pounds of plutonium aboard Galileo and 23.1 pounds aboard Ulysses. NASA's Aerospace Safety Advisory Council had expressed concern about safety precautions to prevent the spread of highly radioactive debris in case of a launch accident.[5]

If either or both missions missed the Jupiter launch window, the next opportunity to launch them would not occur until July 1987. The *Challenger* 51-L mission was thus time critical. It was required to get off the pad and return on a tight schedule so that the orbiter could be reconfigured for Centaur. Another major effort for 1986 was the launch of the Hubble Space Telescope, the most powerful astronomical instrument of the space age, scheduled for October. With that instrument in orbit, scientists hoped to see planets circling other stars and peer farther back in time than ever before, possibly to the Big Bang and the beginning of the universe. This

---

[4] Ulysses and Galileo were to be launched to Jupiter by their attached Centaur rockets after being deployed in low Earth orbit by their shuttles. The launch window was a short period of time (days) when the relative positions of the Earth and Jupiter made it possible for the spacecraft to intercept the big planet in about 600 days. The minimum energy trajectory followed an elliptical path part way around the sun.

[5] After the *Challenger* 51-L accident, the Department of Energy considered a requirement that Ulysses and Galileo be launched only when a west wind was blowing across the launch site so that in the event of a launch accident the radioactivity would be blown out to sea. Both missions later were postponed indefinitely and the spacecraft placed in storage.

was the year that shuttle flights were to be inaugurated from Vandenberg Air Force Base.

With the loss of *Challenger*, NASA's entire launch program for 1986 collapsed. Lost with it was the last chance in this century to observe Halley's comet from the shuttle. A nine-day observing mission, Astro 1, had been planned in early March aboard *Columbia*.

Building Down

As the Presidential Commission discovered, the investigation of the *Challenger* accident had to review the origin and evolution of the shuttle, the world's first nonballistic, manned spacecraft. The vehicle initially was proposed as a unit in a manned interplanetary transportation system by a Space Task Group appointed by President Nixon in 1969. It would shuttle crews and cargo between the ground and low Earth orbit and provide transportation to and from a permanent, habitable station orbiting the Earth. Such a station would serve as a base for lunar and planetary flights by other types of vehicles in the transportation system, leading to a crew landing on Mars in the mid-1980s.

Because of economic and political constraints in the 1970s when the United States was experiencing a time of troubles, the Space Task Group report of 1969 was never quite realized. Only the shuttle materialized. The permanent manned space station was not authorized until 15 years later.

From a long-term point of view, the 51-L disaster was the end product of budget compromises that required NASA to abandon its original design of a fully reusable spaceship and substitute a partly reusable vehicle at about half the development cost. As one critic has said, "they had to build it down to a price, not up to a standard."[6]

In complying with the financial restrictions of the

---

[6] David Webb of the National Commission on Space speaking on the "McNeil-Lehrer News Hour," May 5, 1986.

Office of Management and Budget, NASA designed a vehicle without a launch escape system that had been standard on all previous manned spacecraft. Film and telemetry evidence have shown that the crew compartment of *Challenger* survived the fireball intact after breaking away from the cargo bay, the wings, and other parts of the orbiter. Experts have surmised that members of the crew were alive as the compartment began its plunge into the Atlantic but perished on ocean impact. This highly speculative aspect of the commission's findings will be described in more detail later, but it relates to a question of design. Could the crew have survived the fall if the compartment had been equipped with a parachute descent system, as were the smaller Mercury, Gemini, and Apollo capsules?

A Tale of Two Airplanes

In 1970, the predominant view of the space shuttle was that of a two-stage transport consisting of two airplanes, a booster and an orbiter. Each was powered by rocket and jet engines. An early design depicted a booster the size of the Boeing 747 and an orbiter the size of the Boeing 707. These vehicles would be launched vertically by rocket engines, the smaller riding piggyback on the larger. At an altitude of 25 or 30 miles, the smaller airplane, the orbiter, would separate from the booster and continue accelerating on its rocket engines to orbital velocity. The pilot of the booster airplane would then shut down the rocket engines, execute a turnaround descent maneuver, and start jet engines to fly the ship back to the launch site at Kennedy Space Center. When the orbiter completed its mission in space, the crew would perform a de-orbit maneuver, allowing the ship to reenter the atmosphere. At a predetermined altitude and velocity, the crew would turn on the jet engines for a conventional approach and landing.

There was nothing new in the piggyback concept of dual airplane flight. It had been developed by the British to extend nonstop air service from England to

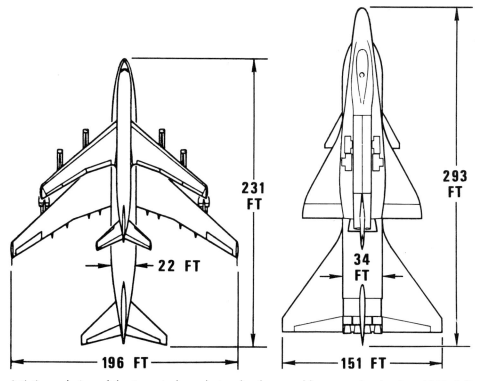

Artist's rendering of the two-airplane design for the reusable space shuttle circa 1969 *(left)* and modification with winged, fly-back booster *(right)* 1970. (NASA)

Egypt during World War II. NASA's predecessor, the National Advisory Committee for Aeronautics, and the U.S. Air Force had experimented with it. In fact, the multistage airplane predated the multistage rocket. In American experience, the Bell X-1 and X-2 and North American Aviation's X-15 were carried aloft by bombers before starting their climb into the stratosphere.

The two-stage airplane design for a resuable space shuttle was already being considered when Neil Armstrong and Edwin E. (Buzz) Aldrin, Jr. landed on the Moon, July 20, 1969. The national euphoria generated by that achievement and the lunar missions that followed it seemed to provide a favorable political climate for a more advanced transport capable at the outset of exploiting the commercial opportunities in Earth orbit and of providing the means for constructing and servicing a space station.

The members of Nixon's Space Task Group were practical men: scientists, engineers, experienced officials. The chairman was the Vice President, Spiro T. Agnew, who, by virtue of his office, also served as chairman of the National Aeronautics and Space Coun-

cil, an advisory group to NASA. Others were Robert C. Seamans, Secretary of the Air Force; Thomas O. Paine, the holdover NASA administrator from the Johnson administration; Lee A. DuBridge, the President's science adviser.

In the afterglow of the manned lunar landings, it seemed reasonable to these men to plan a program of operations on the Moon and an expedition to Mars. But the first step was to create an economically rational system of space transportation. It was evident to them that the Saturn 5–Apollo system would not be developed much further and was not economically viable. It was essentially an ad hoc project designed to put the first men on the Moon and, as a byproduct of this Cold War contest, to conduct limited scientific exploration. It had enabled the nation to conduct the most far-ranging program of exploration in history. It was the boldest exploratory effort by the United States since the naval expedition of Charles Wilkes (1840) that established the continental extent of Antarctica and the most conspicuous technical achievement since the atomic bomb.

In this view of the two-airplane design, the Boeing 707–sized orbiter rides piggyback on the Boeing 747–sized booster. (Source unidentified)

The Space Task Group and the designers who visualized the shuttle regarded it as a space-adapted airplane, capable of maneuvering inside and outside the sensible atmosphere. Its big wing, rudder, and related aerodynamic surfaces allowed controlled flight in the atmosphere but confined it to orbital altitudes of less than 600 miles. This limit was imposed by the speed at which a winged vehicle could safely reenter the atmosphere without breaking up. Because the Apollo command module was shaped like a top and heavily shielded, it could reenter at lunar return velocity. Although the shuttle would have space engines that would provide limited maneuverability in orbit, it could not carry enough fuel to make large changes in altitude. Operationally, the vehicle was to be designed for orbits of less than 300 miles altitude. That constraint would

require upper-stage vehicles for the transfer of crew or cargo to higher orbits or to the Moon. It also dictated the economically optimum altitude for a space station served by the shuttle.

Upper-stage vehicles for transfer to geostationary or lunar orbits were included in the Space Task Group's proposal. It listed a space tug operating from a space station to deliver satellites to geostationary orbit 22,300 miles above the equator. It described an interorbital transfer rocket to deliver cargo or passengers to lunar orbit, where a specialized landing vehicle would complete the trip to the surface.

In broad outline, the task group report has remained the basic space program of the United States, supplemented by relatively short-term or ad hoc scientific missions or missions of opportunity. Although the task group's program, including the landing on Mars, appeared to be as achievable in 1969 as the lunar landing had appeared in 1961, it lacked the competitive motivation engendered by the Cold War that had energized and financed Project Apollo. Having demonstrated technological superiority in space, the United States could not rely on competition with the Soviet Union to provide funds for a continuing adventure on the Moon or anywhere else in space.

By 1970, the nation's priorities were changing. The priority accorded manned space flight in the 1960s had become overwhelmed by new priorities growing out of an array of military and socioeconomic needs. In this new context, if manned space flight was to develop beyond Apollo, it had to be given an economic justification. NASA and its advisers turned to another rationale: cost-effectiveness. The space agency projected the reusable shuttle as the means of reducing space transportation costs and of producing enough income from cargo to pay some operating costs. The shuttle would serve as a common carrier for commercial as well as scientific and military satellites, at operating costs lower than those of expendable or throwaway launch vehicles. Optimistic projections showed that a space shuttle transportation system could become operationally self-supporting, with the possibility of eventually amortizing the nation's investment in it. Although these assumptions were challenged vigorously, they provided a part of the government's rationale for building the shuttle. The other part was the belief that the United States must maintain a manned presence in space, since the USSR was doing so.

In the cost-effectiveness debate, the shuttle's original role as a unit in an interplanetary transportation system was sidetracked, along with the long-term goal of a Mars landing. The Moon was abandoned as the American space establishment retreated to low Earth orbit with the end of Apollo. Still, manned space flight became a target of criticism from several sources—physical scientists who complained that manned missions diverted funds from significant unmanned scientific investigations and social scientists who opposed

the diversion of funds from social amelioration programs and entitlements.

NASA defended the shuttle as a cost-effective substitute for expendable rockets, such as the Delta, Atlas Centaur, and Titan III series. It was depicted as offering the capabilities of both expendable rockets and manned spacecraft: it could launch scientific as well as commercial and military satellites at a potentially lower cost than expendable rockets could launch them, and it could carry a human crew to monitor their deployment and perform experiments and observations as well.

Once having assumed the obligation of seeking cost-effectiveness, NASA's shuttle program became subject to the initial constraint of development cost. Apollo had largely escaped that net, but the shuttle was caught in it, and that constraint was to shape its design.

The original two-airplane design had elicited an imposing response from the aerospace industry. Convair Division of General Dynamics offered a booster airplane 240 feet long that would carry a 174-foot orbiter with a 25-ton cargo capacity to an altitude of 270 nautical miles (310.5 statute miles). Lockheed proposed a 747-sized booster and 707-sized orbiter with similar cargo capacity. Rockwell International offered a 263-foot booster with a 230-foot wingspan and a 202-foot orbiter with a 146-foot wingspan.

These large vehicles were to be powered by rocket engines using hydrogen-oxygen propellant like the upper stages of the Saturn 5. Their size was dictated by the size of their internal propellant tanks, especially in the booster.

Contracts for Phase B studies of the two-airplane shuttle design were awarded by NASA to the McDonnell Douglas Astronautics Company of St. Louis and to North American Rockwell Corporation of Downey, California. In Phase B, the conceptual design (Phase A) was to assume a physical shape with dimensions and cost estimates. The two-airplane shuttle never got beyond Phase B.

Preliminary cost estimates ranged from $10 to $12 billion (1970) over a development period of seven to eight years. Saturn 5–Apollo development had cost the nation an estimated $18 billion in the 1960s. Gone with the 1960s were the urgency and challenge that had conferred a high priority on Apollo. The Office of Management and Budget put a cap of $5.5 billion on development of the shuttle. That sum would not buy the two-airplane shuttle or a fully reusable design. An alternative design costing half as much had to be devised to fit the Procrustean bed of the OMB.

NASA designers and their industrial counterparts understood a hard reality—that such a cut in development cost would inevitably result in an increase in operating cost, either by reducing payload capability or by requiring the use of expendable components. This inverse relationship diminished at the outset any prospect of making the vehicle cost-effective, but still NASA was hopeful that it could produce a vehicle that would undercut launch costs of expendable rocket launchers.

A design breakthrough was achieved by shifting the orbiting vehicle's propellant tank from inside the orbiter to outside it and then dumping the tank in a convenient ocean (it turned out to be the Indian Ocean) after the ascent to orbit was completed. With the expendable external tank, the designers could reduce the length of the orbiter from 166 to 122 feet and double cargo capacity by reducing landing weight. In an early configuration, the new design retained the two airplanes with a 747-sized booster, the 122-foot orbiter, and the external tank.

Estimated development cost was reduced to $8.2 billion. Operating costs were calculated to increase from $2 million to $3.8 million per flight. Replacement of the expendable external tank alone would cost $600,000. However, the executive budget agency found that $8.2 billion was still too high a cost for a nation with a war in Southeast Asia and domestic problems at home.

The next step to bring development cost down was to eliminate the fly-back booster airplane and substitute rockets. These could be made partially reusable by equipping them with parachutes so that after burnout they could be dropped into the sea and after a controlled descent, be recovered and refurbished.

By the end of the first quarter of 1971, the shuttle had been scaled down to a partially reusable, rocket-boosted airplane. It consisted of three parts: the 122-foot orbiter with a 78-foot wingspan and a cargo bay 60 feet long and 15 feet across; the external propellant tank 154.2 feet long and 27.5 feet in diameter, holding 518,000 gallons of liquid hydrogen and oxygen to feed the orbiter's three engines; two solid fuel rockets attached to the tank, each 12.17 feet in diameter and 149.5 feet long. They were the largest solid fuel rockets ever built. The external tank, which some engineers considered proposing for conversion as a space station, was the largest structure to be flown since the Saturn 5 moon rocket. Empty, it weighed 77,000 pounds, later reduced to 66,000 pounds.

The final step in defining the shuttle was to eliminate jet engines from the orbiter. The engines were considered for a time, to allow the orbiter to maneuver in the atmosphere, especially on approach and landing. When the jet engines were taken away, the orbiter assumed its final shape as a rocket boosted glider. Once this machine reached orbital velocity on ascent, the big propellant tank was jettisoned and the orbiter's main engines were silent. The only propulsion that remained was provided by twin orbital maneuvering engines and arrays of forward and aft attitude maneuvering jets. Once the orbital maneuvering engines were fired to allow reentry, the orbiter was committed back to the atmosphere, where it became the world's biggest glider.

For a short time, liquid fuel rocket boosters were considered in lieu of the solid rocket boosters. The liquid fuel rockets were ruled out as more costly than the solids and more difficult to recover and refurbish.

A belief by designers and engineers that the solid rocket boosters represented a mature and safe technology provided the rationale for eliminating a launch escape system, although aircraft-style ejection seats were built into *Columbia* for the commander and pilot, to be used if an emergency developed during the first four "test" flights. As mentioned earlier, they were removed when *Columbia* was refurbished in 1983. Despite the

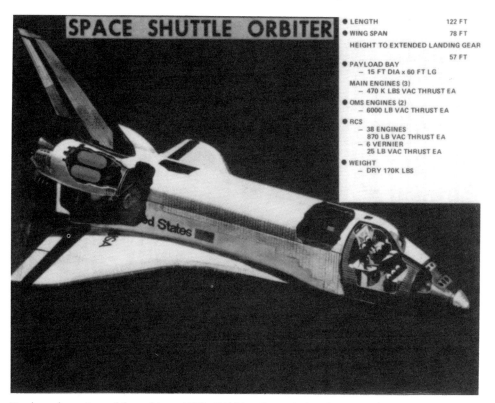

SPACE SHUTTLE ORBITER

- LENGTH   122 FT
- WING SPAN   78 FT
- HEIGHT TO EXTENDED LANDING GEAR
               57 FT
- PAYLOAD BAY
  - 15 FT DIA x 60 FT LG
  - MAIN ENGINES (3)
  - 470 K LBS VAC THRUST EA
- OMS ENGINES (2)
  - 6000 LB VAC THRUST EA
- RCS
  - 38 ENGINES
    870 LB VAC THRUST EA
  - 6 VERNIER
    25 LB VAC THRUST EA
- WEIGHT
  - DRY 170K LBS

Final configuration of the orbiter, 1972. (NASA—*Commission Report,* vol. 2)

ejection seats, launch escape or bail-out from the orbiter during the two minutes that the solid rocket boosters were firing were deemed nonsurvivable. Emergency landing and mission abort, possible only after solid rocket booster burnout, were survivable only under specific conditions that allowed the orbiter to glide back to the launch site, to a transatlantic landing site, or around the Earth to a landing at Edwards Air Force Base, California.

President Nixon authorized the construction of the Space Shuttle Transportation System on January 5, 1972. The estimate for design, development, testing, and evaluation of two vehicles (*Enterprise and Columbia*) was $5.15 billion. The total cost of manufacturing a fleet of five vehicles was pegged at $8.1 billion. Initial development began under a new NASA administrator, James C. Fletcher, an aerospace industry executive who had become president of the University of Utah. He succeeded Thomas O. Paine, a General Electric Company executive, who had headed the space agency during most of the Apollo operations era. NASA administrators, like cabinet officers, were political appointees of the President.

The Boosters

The shuttle's solid rocket boosters (SRBs) provided 71 percent of the total thrust at liftoff during the first two minutes (or first stage) of the ascent, when the shuttle reached an altitude of about 27.6 miles. The motors were manufactured by Morton Thiokol, Inc., Wasatch Division, of Brigham City, Utah; the rocket structure, by the McDonnell Douglas Astronautics Company, Huntington Beach, California; the parachutes, by the Pioneer Parachute Company of Manchester, Connecticut; and checkout, assembly, launch, and refurbishment, by United Space Boosters, Inc., Sunnyvale, California.

Solid fuel rocket technology has a long history, beginning with gunpowder. The modern solid fuel motor (SRM) originated with the invention of a polysulfide rubber called thiokol in 1927. At the end of 1945, C. E. Bartley, an engineer at the Jet Propulsion Laboratory of the California Institute of Technology, discovered that this material, which was manufactured for several industrial uses by the Thiokol Chemical Corporation, was a superior rocket propellant. It worked better than

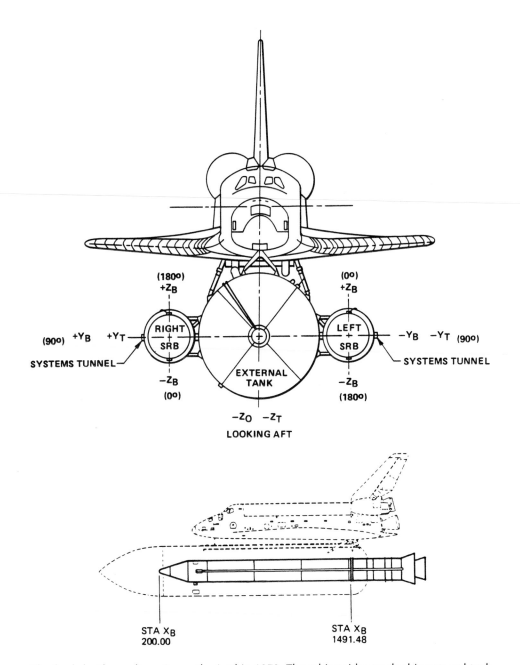

The final shuttle configuration authorized in 1972. The orbiter rides on the big external tank which feeds propellant (hydrogen and oxygen) to the orbiter main engines. Solid rocket boosters are attached to each side of the tank. They provide most of the thrust at liftoff and drop off after two minutes. The orbiter engines continue firing until 8 minutes 34 seconds after liftoff, when they are shut down and the big propellant tank is dropped. The orbital maneuvering (space) engines, shown in the next drawing, are then fired to reach final orbit. (NASA)

1 Orbital Maneuvering System

Two engines
  Thrust level = 6,000 pounds each

Propellants
  Monomethyl hydrazine (fuel) and
  nitrogen tetroxide (oxidizer)

2 Reaction Control System

One forward module, two aft pods

38 primary thrusters (14 forward, 12 per aft pod)
  Thrust level = 870 pounds each

Six vernier thrusters (two forward, four aft)
  Thrust level = 25 pounds each

Propellants
  Monomethyl hydrazine (fuel) and
  nitrogen tetroxide (oxidizer)

3 Main Propulsion

Three engines
  Thrust level = 375,000 pounds each

Propellants
  Liquid hydrogen (fuel) and
  liquid oxygen (oxidizer)

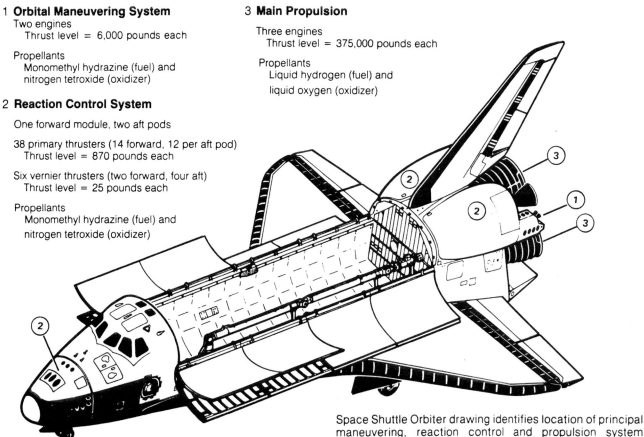

Space Shuttle Orbiter drawing identifies location of principal maneuvering, reaction control and propulsion system engines.

The orbital maneuvering engines. (NASA)

asphalt, which JPL had developed as a solid fuel propellant in JATO rockets for aircraft and racing cars.[7]

In the late 1940s, JPL researchers built a solid propellant motor by bonding thiokol to the rocket case, leaving a hollow channel running down the center shaped like an 11-point star. This channel shape allowed the fuel to burn evenly and provide maximum thrust at liftoff. When the star points burned away, the center hole became rounded, and initial launch thrust was reduced.[8]

JPL used this fuel configuration in the Sergeant battlefield missile the laboratory developed for the U.S. Army in the 1950s. Scaled-down Sergeant rockets com-

prised the upper stages of the Jupiter C missile that put the first American satellite, Explorer I, into orbit, January 31, 1958. It was JPL's solid rocket research that "constituted the fundamental breakthrough that made possible the massive solid rockets of the missile and space eras," according to the JPL historian Clayton R. Koppes.

The shuttle SRBs, like the Minuteman, Polaris, and Poseidon missiles and the Titan III series of space launchers, are the products of this 30-year-old technology. In the light of their genealogy, the SRBs represented a mature techology that had been thoroughly tested and, presumably, understood. The Thiokol Chemical Corporation having won the SRB motor manufacturing contract in 1973, was merged with the Morton Salt Company in 1982. The propellant is a more complex material today than the early mix. It consists of an atomized aluminum powder as fuel (16 percent); ammonium perchlorate as oxidizer (69.93 percent); oxi-

[7] See Clayton R. Koppes, *JPL and the American Space Program* (New Haven: Yale University Press, 1982).

[8] In the shuttle SRB, initial launch thrust was reduced at 55 seconds by this fuel configuration to avoid overstressing the vehicle at Max Q.

This segment of the left-hand solid rocket booster was scheduled for mission 61-G, which never flew. The booster was taken apart in the Vehicle Assembly Building in April 1986 to determine if assembly procedures might have contributed to the 51-L accident. The solid rocket motor consists of four such casting segments, each about 24 feet long and 146 inches in diameter. Each in turn is made up of two segments which are joined by a factory joint (center ring). The center joint is covered with insulation installed at the factory around the propellant and a liner. The end joints are called field joints because they are made at the Kennedy Space Center, where the four casting segments are shipped by rail from Utah. The field joints are metal and are sealed with two Viton (rubber) O-rings and putty, and the segments are mated with 177 pins around the circumference of the steel case. Each field joint consists of a tang and clevis to which thin metal shims are added to ensure a tight fit and squeeze on the rubber O-rings. (NASA)

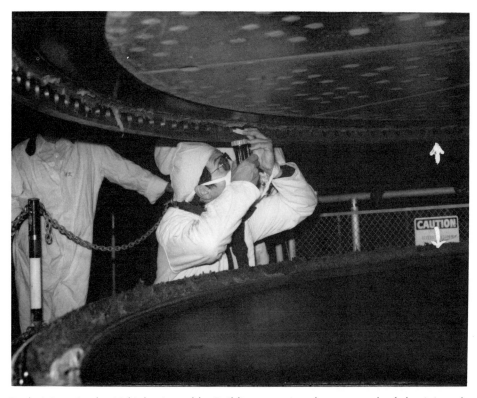

Technicians in the Vehicle Assembly Building examine the tang end of the joint of a disassembled segment of the 61-G left booster *(upper arrow)*. Strands of putty can be seen clinging to the metal. Below is the clevis end of the joint with the putty laid on. The O-ring seals lie behind the putty. (NASA)

dized iron powder as a catalyst (0.07 percent) and polybutadiene acrylic acid acrylonitrile as a rubber-based binder (14 percent) that has supplanted the older polysulfide rubber (thiokol) binder.

At the Morton Thiokol factory, the propellant mix was loaded into four motor segments. When hardened, it had the consistency of a sponge rubber eraser and was easily storable. The segments were then shipped by rail to the Kennedy Space Center in Florida where they were stacked vertically in the Vehicle Assembly Building by a processing team to form the 149.5-foot high booster.

The segments were attached by a tang and clevis joint fastened with 177 steel pins. When the motor segments were joined, there was a gap between the propellant mass and the internal motor case insulation where the segments met. Heat-resistant zinc chromate putty was applied in one-eighth inch strips to to fill the gap. The clevis joint was then sealed by a pair of synthetic rubber O-rings which were installed in grooves around the circumference of the motor case. They were

insulated by the putty against erosion by hot gas when the motor fired.

As the propellant was ignited (electrically), internal gas pressure was expected by the designers of this sealing system to force the primary O-ring to seat in its groove to seal the gap between the propellant and the motor case insulation at the joint. If the primary O-ring failed to seat properly, allowing hot gas by blow by it, the secondary O-ring, downstream of the primary, was expected to prevent the hot gas from reaching and penetrating the one-half-inch thick steel motor case.

The design of the SRB joint sealing system was similar to that developed for the Air Force in the solid rocket boosters of the Titan III launcher, but was more conservative in that it had two O-rings. The Titan boosters had only one. To "man rate" the shuttle, a secondary O-ring in the shuttle SRM was designed to meet a safety requirement of redundancy.

Before the *Challenger* accident, it was believed by some shuttle observers, including this writer, that the

Technicians examine the clevis end of the joint of the disassembled 61-G segment. (NASA)

SRBs were as safe and reliable as rockets could be. In the words of the authoritative journal *Aviation Week and Space Technology*, the design of the solid rocket motor "used conservative specification margins and proved technology to eliminate the possibility of failure of the motor case, its joints or the propellant itself."[9]

[9] "Shuttle 51-L Loss," *Aviation Week and Space Technology*, February 10, 1986, p. 53.

It was the liquid fuel main engine system in the orbiter that had been wracked with development problems during the late 1970s. By contrast, the SRBs had reportedly passed their static tests successfully and were considered to be problem free.

The perception that the SRBs were virtually foolproof was widely shared in NASA during shuttle development and was strongly communicated to the news media correspondents. The boosters and their operation

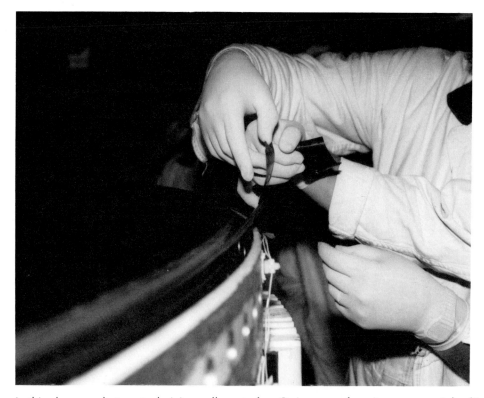

In this close-up photo, a technician pulls part of an O-ring away from its groove or "gland" with strip of tape to look for imperfections. The two O-rings, primary and secondary, line the entire circumference of the motor joint (458.68 inches). Each is 0.28 inches in diameter and is made in five lengths glued together. (NASA)

were initially described to reporters on October 14, 1980 by George Hardy, the booster manager at the Marshall Space Flight Center.

Contrary to the development of the main engines, he said, the emphasis on solid rocket boosters "has been to maximize the state-of-experience on solid booster systems from the past and minimize pushing the state of the art. Now, the reason for that is quite clearly to minimize the ground test data base that would otherwise be required to man rate the solid propulsion system."[10]

During first-stage boost, Hardy said, the shuttle would reach velocity of about 4,400 feet per second and an altitude of 24 nautical (27.6 statute) miles. From ignition to burnout would be about 2 minutes, he said. Then on signal from the orbiter, the boosters would be separated (from the external tank) at forward and aft attachment points by pyrotechnic devices. Small sepa-

ration motors would provide the push to clear the boosters from the tank and the orbiter riding on it.

The boosters would continue coasting to an apogee of 225,000 feet before starting to fall back to the ocean. At 16,000 feet, an altitude switch would signal the ejection of the nose cap on each booster and the deployment of a pilot parachute. This chute would deploy a drogue chute about 50 feet in diameter. At 6,500 feet, a cluster of three parachutes each 105 feet in diameter would open and slow descent so that each booster would strike the water tail first at 85 feet per second or 60 miles an hour. The boosters would then float, buoyed by trapped air, until retrieved by ships.

Hardly described the SRB motor case as a scaled-up version of the Titan III motor case. It was built of the same materials, with the same basic processes, and in fact used the same contractors and subcontractors, he said. There was nothing new about the SRB.

The propellant, Hardy said, had "an extensive experience base." The internal insulation was predomi-

[10] Space Transportation System Briefing Series no. 3, Marshall Space Flight Center, October 14, 1980.

Looking down on a mission 61-G segment, the camera shows how the propellant (PBAN or polybutadience acrilonitrile) is cast in the segment with an inhibitor called NBR (nitrile butadience rubber). The center ''hole'' or core is known as the bore. At ignition, the initial burning starts in the bore and burns inside out. (NASA)

nantly rubber and was typical of the internal insulation used in the vast majority of all the solid rocket motors built to date. There were approximately 9 tons of insulation in a single motor, he said, adding: ''We do have a factor of safety on that internal insulation between 1.5 and 2 at various specified locations within the motor case.'' The motor cases were planned for 20 uses.

During this briefing, Hardy displayed a picture of a full-scale solid rocket motor, one of seven, he said, that were tested in the desert of northern Utah, starting in 1977. ''There were very few and very minor changes that had to be made in the motor design as a result of the early tests,'' he said. Ground tests of these types are rather expensive, and so ''one of the specific design

objectives from the beginning has been to utilize the very conservative factors in safety—utilize proven, well-known processes, techniques, and materials. And that's so we build the confidence that was required to man-rate the solid rocket booster with such a limited number of tests.''

The Achilles Heel

To those who shared the views presented in this briefing as an article of faith, the apparent failure of the seal on the right booster that destroyed *Challenger* was incredible. But it was not incredible to a number of

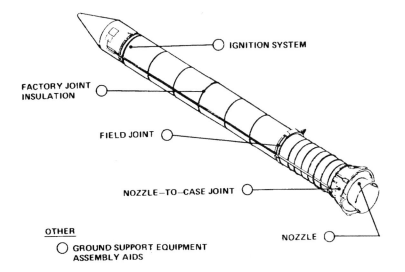

The joints linking the segments assembled at the Kennedy Space Center are known as "field" joints. Portions of segments assembled at the factory are called "factory" joints and are insulated there. The leak that resulted in the destruction of *Challenger* occurred at a field joint between the aft-center and aft segments. (NASA)

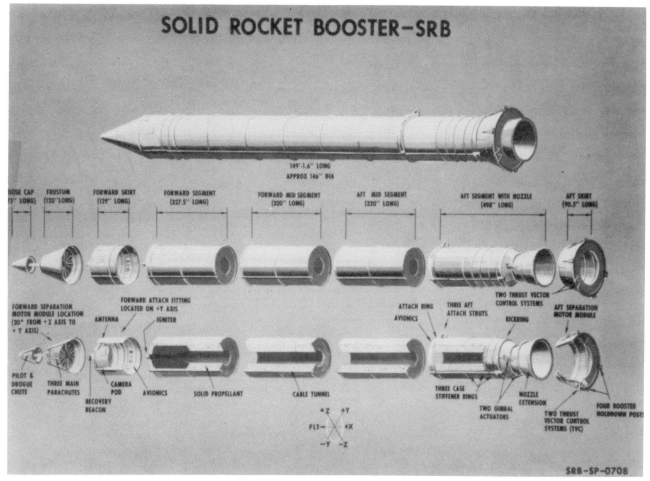

Design of the solid rocket booster. (*Commission Report*, vol. 2)

Structure of the field joint showing tang, clevis, and O-rings. (*Commission Report*, vol. 1, p. 57)

engineers at Marshall and at Morton Thiokol who were aware of a long history of seal problems, known only to insiders and regarded with increasing anxiety as safety concerns.

NASA had let the cost-plus, award-fee contract valued at $800 million to manufacture solid rocket boosters to Thiokol Chemical Corporation, November 20, 1973. Competing for the contract were three other rocket manufacturers: Aerojet Solid Propulsion Company, Lockheed Propulsion Company, and United Technologies. The Source Evaluation Board rated Thiokol fourth on design, development, and verification, according to the commission report, second on manufacturing, refurbishing, and product-support capability, and first on management. Aerojet was first in the category of overall mission suitability, and Thiokol was tied with United Technologies for second place.

In its selection report, NASA headquarters stated that Thiokol showed substantial cost advantages. That was a key consideration in a period when the development of the shuttle had to fit the OMB's allowance.

Thiokol began testing its motors the following year. NASA proclaimed that the tests were successful or had accomplished their objectives. In contrast to the repeated test failures of the orbiter main engines, which threatened to generate a scandal, Thiokol's solid rocket motors acquired a legend of infallibility.

However, engineers at the Marshall Space Flight Center, where supervision of SRB development was vested, discerned that the motor design had an Achilles heel when Thiokol tested the strength of the steel rocket case in 1977. The "hydroburst" test simulated a motor firing by pressurizing the case to 1,500 pounds per square inch. This pressure was one and one-half times that exerted by the motor after ignition.

The hydroburst test proved the strength of the case but also revealed a sinister pressure effect. Under pressure, the tang and inner lip of the clevis in the field joints where two segments were mated bent away from each other instead of toward each other as expected.[11] This effect, called joint rotation, opened a gap that allowed the secondary O-ring to become unseated.

As Arnold Thompson, a Thiokol supervisor, related: "We discovered that the joint was opening rather than closing as our original analysis had indicated and, in fact, it was quite a bit. I think it was up to 52 thousandths of an inch at that time . . ."[12]

---

[11] Three field joints were made when the rocket was assembled at the Kennedy Space Center.

[12] Testimony before the Presidential Commission, February, 25, 1986. *Report of the Presidential Commission on the Space Shuttle Challenger Accident* (Washington, D.C., 1986), vol. 5, p. 1435.

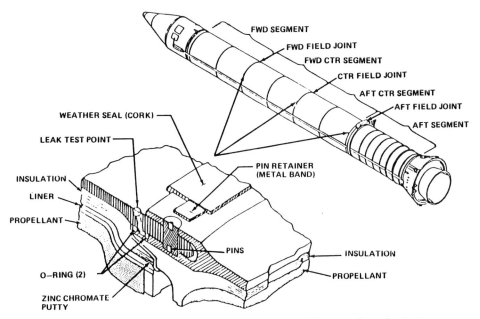

Location of the field joints. At left, design of joint, showing the tang-clevis fit, O-rings, putty, and leak check point. (NASA—*Commission Report,* vol. 2, p. L-40)

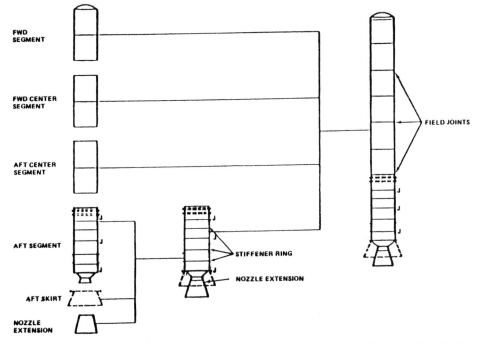

Segments of the solid rocket motor (SRM) and field joints. (Morton Thiokol—*Commission Report,* vol. 2, p. H-628)

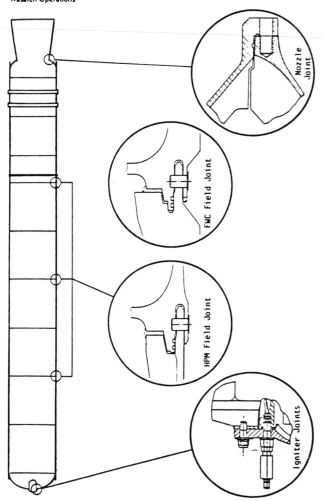

MORTON THIOKOL INC.
Wasatch Operations

Detail of the current high-performance motor field joint *(lower center circle)* compared with the advanced filament-wound case field joint *(upper center)* projected for later use. Sketches of nozzle joint *(top)* and igniter joint *(bottom)*. (Morton Thiokol— *Commission Report,* vol. 2)

Thiokol reported this test effect to the NASA shuttle program office at Marshall, according to the commission report. It added that Thiokol engineers did not believe that joint rotation would cause problems and did not schedule additional tests to confirm or disprove the joint gap effect.

Marshall engineers were concerned about the gap effect, however, and began to complain about it in the fall of 1977. The commission cited a memorandum of September 2, 1977 from Glenn Eudy, chief engineer in Marshall's solid rocket motor division, stating that the Thiokol design allowed O-ring clearance. He told Alex McCool, director of the Marshall structures and propulsion laboratory: "some people believe this design deficiency must be corrected by some method, perhaps design modification." He described O-ring clearance as a "very critical solid rocket motor issue."

Another memorandum was submitted by Leon Ray, Marshall engineer on October 21, 1977 characterizing the Thiokol joint design as "unacceptable." He recommended redesign of the tang and reduction of tolerance on the clevis for a "long term fix." The tang, he said, could move outboard and "cause excessive joint clearance resulting in seal leakage."

In still another memorandum dated January 9, 1978, John Q. Miller, chief propulsion engineer at Marshall, recommended rejection of certain proposals from Thiokol that standards be relaxed for clevis joint acceptance "because of excessive deviation from military standard requirements." A year later, in a memorandum January 19, 1979, Miller repeated the recommendation, adding more succinctly: "We find the Thiokol position regarding design adequacy of the clevis joint to be completely unacceptable for the following reasons: (1) the gap created by excess tang-clevis movement causes the primary O-ring seal to function in a way that violates industry and government O-ring application practices; (2) excessive tang-clevis movement allows the secondary O-ring to become completely disengaged from its sealing surface on the tang." The clevis joint's secondary O-ring seal "has been verified by tests to be unsatisfactory," he concluded.

These memoranda disclosed that Marshall engineers diagnosed a flaw in the motor joint seals four years before the shuttle began to fly, but the commission did not find any record that any change was made in the seal design.

In 1980, a Space Shuttle Verification-Certification

SRM-HPM Field Joint

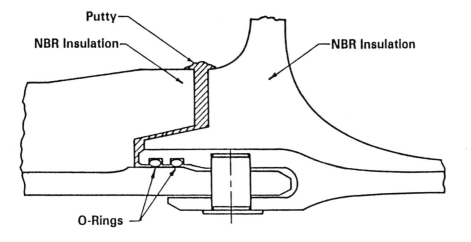

MORTON THIOKOL, INC.
Wasatch Division

'NFORMATION ON THIS PAGE WAS PREPARED TO SUPPORT AN ORAL PRESENTATION

High-performance motor field joint showing tang-clevis junction, location of primary and secondary O-rings, putty and factory (NBR) insulation. (Morton Thiokol—*Commission Report*, vol. 2)

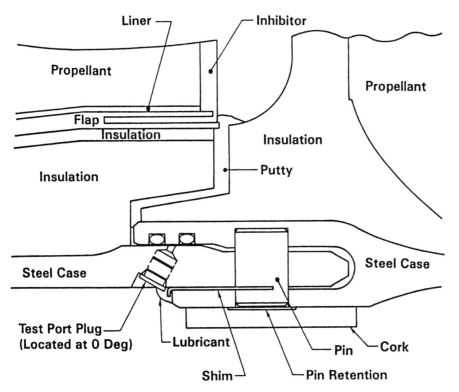

Field joint showing relationship of propellant, insulation, putty, and leak test port. (Morton Thiokol—*Commission Report,* vol. 5)

The 149.16-foot-tall boosters are attached to each side of the 154.2-foot external tank in the huge Vehicle Assembly Building at the Kennedy Space Center. On the launch pad, each booster is attached to the mobile launch platform at the aft skirt by four bolts that are severed by small explosives at liftoff. (NASA)

Committee was formed to study the flightworthiness of the shuttle transportation system, including the orbiter, the external tank, and the boosters. The propulsion subcommittee raised several caveats about the joint seal design. One referred to a test to find out if the joint leaked after the booster was assembled. The leak test was done by forcing gas under pressure into the joint to determine whether the primary O-ring was actually sealing the joint. The criticism was that the leak test moved the primary O-ring in the wrong direction, that is, away from the groove to be sealed. However, both Marshall and Thiokol engineers argued that motor pressure at ignition would move the primary O-ring into its proper position for sealing the joint. The subcommittee's second caveat warned that the putty laid over the joint provided uncertain protection against the burning of the O-rings by hot gas after ignition. Both issues were to grow in magnitude as the shuttle began to fly.

The propulsion group characterized the redundancy of the secondary O-ring seal as a verification concern. But its report stated that it understood that the main purpose of the secondary O-ring was to test the primary O-ring and that "redundancy is not a requirement." When the commission asked him about this, George B. Hardy, then Marshall deputy director for science and engineering, said the belief that redundancy was not a requirement was a misunderstanding. He said he did not know where the group got that idea.

The Verification-Certification Committee specifically called for firing tests to verify field joint integrity after the booster segments were stacked in their flight configuration. During a burst test, one of the O-rings failed and leaked. The committee considered the test inadequate because the analysis of the results did not cover the extent to which the gap in the joint was opened at pressurization or by bending loads (that would be imposed during launch).

NASA's response to this criticism held that the

Aft segments and skirts of the boosters are shown in this photo as assembly proceeds. The entire weight of the shuttle rests on the aft skirts, through which motor nozzles protrude. (NASA)

original hydroburst tests had satisfied the intent of the committee's recommendations.[13] "NASA specialists have reviewed the field joint design, updated with larger O-rings and thicker shims, and found the safety factors to be adequate for the current design," the response stated.[14]

[13] *Commission Report,* vol. 1, p. 124.
[14] *Commission Report,* vol. 1, p 149: "SRM Program Response," NASA August 15, 1980.

The motor was certified by the propulsion group on September 15, 1980, and the joint was classified on the shuttle critical items list as 1-R, or redundant. However, the classification carried a caveat. It stated that redundancy of the secondary field joint seal could not be verified after pressure in the motor case reached 40 percent of maximum operating pressure. The reason: "It is known that joint rotation occurring at this pressure

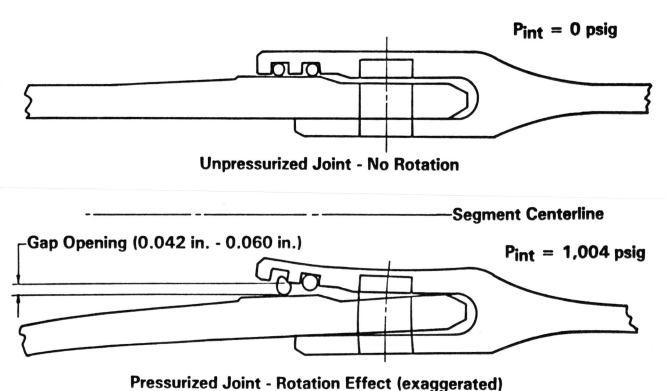

$P_{int} = 0$ psig

Unpressurized Joint - No Rotation

Segment Centerline

Gap Opening (0.042 in. - 0.060 in.)

$P_{int} = 1,004$ psig

Pressurized Joint - Rotation Effect (exaggerated)

Gas pressure generated by ignition of the solid rocket motor opens a gap in the field joint between tang and clevis, preventing the seals from functioning properly and allowing gas leakage. The effect was called joint rotation. (NASA)

level with a resulting enlarged extrusion gap causes the secondary O-ring to lose compression as a seal. It is not known if the secondary O-ring would successfully re-seat if the primary O-ring should fail after motor case pressure reaches or exceeds 40 per cent of maximum expected operating pressures."[15]

The Secret That Never Leaked

The Achilles heel of the shuttle system showed up on STS-2, the second test flight of *Columbia*, during its launch, November 12, 1981. When the boosters were recovered, inspectors found that hot gas had penetrated the putty and damaged the primary O-ring in the aft field joint of the right booster. At one point, the ring was eroded 0.053 inches or 19 percent of its thickness.[16]

[15] NASA 1980, critical items list, *Commission Report*, vol. 1, p. 125)
[16] "Erosion of SRM Pressure Seals," Thiokol report. August 19, 1985, *Commission Report*, vol. 1.

It was one of the worst cases of seal damage in the shuttle program.

The commission found that this anomaly was not reported in the Level 1 (headquarters) flight readiness review for *Columbia's* third test flight and was not reported to the Marshall Space Flight Center's problem assessment system.

Following high-pressure O-ring tests in May 1982, shuttle management at Marshall concluded that Thiokol's dual O-rings did not provide a fully redundant system because the secondary O-ring would not always function after joint rotation following ignition. It para-phrased the conclusion reached five years earlier by the Marshall engineers who were critical of the Thiokol seal design.

At this juncture, Marshall officially declared that the field joint sealing system was not redundant and could fail with catastrophic consequences. It did not meet the failsafe definition of the Space Shuttle Program Requirements Document, Level 1, of June 30, 1977.

The criticality 1-R designation was changed at Marshall on December 17, 1982, and the O-ring seals of the field joints were reclassified as criticality 1.

The SRB critical items list defined criticality 1 items as those subject to a single-point failure. Leakage of the primary O-ring was classified as a single-point failure "due to possibility of loss of sealing at the secondary O-ring because of joint rotation after motor pressurization." The list summarized the effects of the failure as "loss of mission, vehicle and crew due to metal erosion, burn through and probable case burst resulting in fire and deflagration."

The shuttle criticality 1 list covered 748 items for which there was no backup or redundancy. Only the primary structure and thermal protection system were exempted. The critical items list compiled by the National Space Transportation System and made public March 17, 1986 listed 335 items in the orbiter that were subject to a single-point failure. Most were simple items of hardware. On the solid rocket boosters there were 114 criticality 1 items, of which 59, including the O-ring primary seals, had been granted waivers.

In substance, a waiver meant that a criticality 1 item could be tolerated as a flight risk. In any event, the shuttle was flown with hundreds of such items whether waivered or not.

A rationale for flying the shuttle with a single functional O-ring was written by Howard McIntosh of Thiokol engineering in December 1982. He cited the test history of the Titan booster, which had functioned well, he said, with a single O-ring at the joints. He contended that the tang-to-clevis motion during motor operating pressure would not unseat the secondary seal.[17]

Leon Ray of Marshall, in his 1977 Solid Rocket Motor Joint Leakage Study, had asserted that the tang could move outboard and cause excessive joint clearance resulting in leakage past the seal. When the commission asked him about this on May 2, 1986, Ray said: "When the joint was first designed, the analysis provided by Thiokol said the joint would close, the extru-

sion gap would actually close. We had quite a debate about that until we did a test on the first couple of segments that we received from the manufacturer, which in fact showed that the joint did open. Later on we did some tests with the structural test article. . . . At that time, we really nailed it down."

Ray and John Miller disputed with the Thiokol calculations, and the argument went on for years. It was still in progress when *Challenger* blew up.

Notwithstanding the reclassification from 1-R to 1, Marshall solid rocket motor managers believed, with McIntosh, that the secondary O-ring seal actually was redundant despite joint rotation in all but exceptional cases. Their arguments, made before the commission, will be presented later, but they illustrated how the engineers differed on a technical point, even one with life-and-death implications.

L. Michael Weeks, NASA associate administrator, approved a waiver on the criticality 1 joint, March 28, 1983. That settled the issue of whether the shuttle should continue flying in this condition. He told the commission that he signed the waiver because "We felt at the time, all of the people in the program, I think, felt that this solid rocket motor in particular was probably one of the least worrisome things we had in the program."[18]

Nevertheless, secondary seal redundancy was demonstrated at a nozzle joint after the *Challenger* 51-B mission was launched, April 29, 1985. Primary O-ring erosion was found at the nozzle joint of both right and left boosters. There was blow-by of soot and heat damage to the secondary O-ring in the left booster nozzle joint, but it held and saved the day.[19]

Of prime importance was the discovery that the left booster primary O-ring was burned through in several places and never sealed. Fortunately, the secondary O-ring held the hot gas despite damage. The Marshall Space Flight Center reacted to the nozzle joint primary O-ring failure by imposing a launch constraint on all

[17] *Commission Report*, vol. 1, p. 126.

[18] Commission hearing, May 2, 1986.

[19] The nozzle joint O-ring system was not reclassified as the field joint rings were.

future flights until the cause was analyzed and dealt with. As will be described later, center management believed it had corrected the problem by changing leak check procedure. Consequently, the constraint was waived for each subsequent flight, and the momentum of the launch schedule continued.

The incidence of O-ring damage, including erosion, soot blow-by, and burning of O-ring surfaces, increased each year after 1981, except in 1982 when *Columbia* was launched three times without evidence of seal damage. However, the boosters on STS-4, launched June 27, 1982, were lost when their parachutes failed to open and they plunged into the Atlantic Ocean and sank.

In the six years of shuttle flight operations, evidence of O-ring erosion and blow-by of soot was found in 15 of the 25 shuttle launches, including *Challenger* 51-L. When STS-4 is taken into account, the figures are 15 of 24 launches from which the boosters or parts of them were recovered. From those figures, joint seal damage occurred in 63 percent of the shuttle launches.

In 1981, seal damage showed up in one of two launches; in 1983, in one of three launches; in 1984, in three of five launches; in 1985, in eight of nine launches; and in 1986, in two of two launches. This progression plainly pointed to a flaw in the solid rocket booster sealing system. But although Marshall and contractor engineers became concerned enough about it by summer 1985 to propose various fixes, no effective action was taken. Even though erosion was obviously increasing, NASA flight readiness reviews show that it was rationalized as tolerable because nothing terrible had happened.

An astonishing aspect of this situation was that so far as the public was concerned, it was one of the best-kept secrets of the space age. The documents describing it were not classified and did not need to be. They were buried in the files at NASA headquarters in Washington and the Marshall Space Flight Center in Huntsville, Alabama.

Along with the general public, the astronauts who were flying the shuttle were unaware of the escalating danger of joint seal failure. So were the congressional committees charged with overseeing the shuttle program.

NASA never told them that the shuttle had a problem.

4. Erosion

Following the discovery that erosion had occurred in a booster field joint seal on the second launch of *Columbia*, the effect was seen intermittently and then with increasing frequency as the shuttle entered its so-called operational phase after *Columbia*'s STS-4 mission.[1] Shuttle managers at the Marshall Space Flight Center deemed erosion to be an acceptable risk.

Hot gas paths through the putty and indications of heat effect on the primary O-rings in the right and left booster nozzles were detected when the boosters were recovered from the launch of STS-6, the first flight of *Challenger*, April 4, 1983. Primary O-ring erosion was seen also in the nozzle joint of a qualification test motor fired at Thiokol's Utah test site, March 21, 1983.

As the flight program picked up speed, signs of joint seal damage became more marked and more common. In the postflight inspection of 41-B, a *Challenger* mission launched February 3, 1984, erosion was found on the primary O-rings of the left booster forward field joint and the right booster nozzle joint. The information was given to the Marshall problem assessment system, which analyzed it and reported that no remedial action was required.

The analysis found that the primary O-ring seal in the forward field joint showed a charred area 1 inch long, 0.03–0.05 inches deep, and 0.1 inch wide. The official report said that the "possibility exists for some O-ring erosion on future flights." In this and other

records obtained by the Presidential Commission, it becomes apparent that O-ring damage was being equated with normal wear and tear.

A rationale for accepting O-ring damage was supported by a Marshall analysis purporting to show that maximum possible erosion was 0.09 inches. The analysis cited a laboratory test showing that an O-ring with a simulated erosion depth of 0.095 inches would seal under pressure of 3,000 pounds per square inch. This finding was cited in the flight readiness review for the thirteenth shuttle mission, 41-G, a *Challenger* flight launched October 5, 1984. It supported the official position that erosion "was not a constraint to future launches."

The erosion of the seals of 41-B had been considered at Level 1 (headquarters) during the flight readiness review for the next *Challenger* mission, 41-C, launched April 6, 1984. At that time, it was presented merely as a "technical issue." There is no sense of impending danger in the documents.[2] The commission report stated: "A recommendation to fly 41-C was approved by Level 1 'accepting the possibility of some O-ring erosion due to hot gas impingement.' "

Nevertheless, there were officials at NASA headquarters who did not share the wear-and-tear attitude that could construed from these documents. An outgrowth of the flight readiness review for mission 41-C was a directive issued April 5, 1984, the day before the

---

[1] *Columbia* was declared officially to be "operational" after completing its fourth test flight, July 4, 1982.

[2] NASA flight readiness reviews obtained by the Presidential Commission.

Six years earlier (December 29, 1980), *Columbia* was moved toward pad 39A to be prepared for the first test flight of the shuttle. It was launched April 12, 1981. (NASA)

launch, by NASA Deputy Administrator Hans Mark to Lawrence B. Mulloy, the solid rocket booster project manager at Marshall. It asked Mulloy to conduct a formal review of the field joint sealing process.

Dr. Mark, a former director of NASA's Ames Research Center in California, was not alone in expressing concern about the erosion effect. Earlier in the year, Lieutenant General James A. Abrahamson, associate administrator for space flight, called upon William R. Lucas, the director of the Marshall Space Flight Center, to develop a plan of action to improve the solid rocket motors. This request was formalized in a letter of January 18, 1984. Abrahamson said that NASA was flying motors wherein basic design and test results were not well understood. It was a prophetic observation.

In response to the directive from Hans Mark, Mulloy and his deputy, Lawrence Wear, asked Morton Thiokol for a formal review of the sealing process in field and nozzle joints. The review was to determine the cause of erosion and whether erosion seen thus far was acceptable.

Company engineers replied on May 4 with a plan to increase the protection of the primary O-ring seals by investigating several areas of concern: the pressure used to make leak checks during assembly of the boosters, the loads imposed on the field joints during assembly, eccentricity (deviation from roundness) of the segment cases at assembly, and the quality and method of installation (lay-up) of the putty.

The response had been drawn up by Brian Russell, Morton Thiokol's manager of systems engineering. The commission regarded it as a mere proposal, however. The final response to the directive from Marshall, the commission stated, was not completed until August 19, 1985, fifteen months later. At that time, Thiokol engineers briefed NASA headquarters on solid rocket motor seal problems.

In the meantime, the record of O-ring damage accumulated in shuttle operations included *Challenger* 41-C, April 6, 1984; *Discovery* 41-D, August 30, 1984; *Discovery* 51-C, January 24, 1985; *Discovery* 51-D, April 12, 1985; *Challenger* 51-B, April 29, 1985; *Discovery* 51-G, June 17, 1985, and *Challenger* 51-F, July 29, 1985.

After the headquarters briefing, seal damage continued to be found from the launches of *Discovery* 51-I, August 27, 1985; *Challenger* 51-A, October 30, 1985; *Atlantis* 61-B, November 27, 1985; *Columbia* 61-C, January 12, 1986; and *Challenger* 51-L, January 28, 1986. Marshall and Thiokol continued to look for causes.

The Putty Predicament

At Marshall, John Q. Miller, head of the solid rocket motor section, had sent a warning to George Hardy that failure of the putty to "provide a thermal barrier [to the

O-rings] could lead to burning both O-rings and subsequent catastrophic failure." He said that the barrier effect of the putty was sensitive to temperature and humidity, and that the O-ring leak check procedure might displace putty from a position where it would be effective as a barrier. Miller asserted that the thermal design of the motor joints depended on the thermal protection of the O-rings by the putty.[3]

Morton Thiokol engineers also suspected that problems with the putty were a factor in O-ring erosion. As early as April 1983, the company concluded that the erosion seen on the primary O-ring in the aft field joint of the STS-2 right booster was caused by blowholes in the putty.[4] The holes provided a pathway through which hot gases could impinge on O-rings.

At Morton Thiokol, Brian Russell advised Marshall that the putty blowholes were caused by an increase in leak check pressure that Marshall had adopted in 1984, starting with the launch of *Challenger* 41-B, February 3.

The leak check was made at each joint through a small valve into which nitrogen gas was forced at pressures of 50 to 100 pounds per square inch. Containment of the pressure was measured by gauges. Because there was concern at Marshall that these pressures were too low to test the O-rings and could be held by the putty alone, the test pressure was raised from 50 to 100 psi in nozzle joints and from 50 and 100 to 200 psi in field joints—where secondary O-rings were more likely to become unseated by joint rotation than in the stiffer nozzle joints. The nozzle joints had remained classified as criticality 1-R. Russell believed that the higher test pressures resulted in more erosion by punching holes in the putty.

The commission noted: "This hypothesis that O-ring erosion is related to putty blow holes is substanti-

[3] Routing slip, Miller to Hardy, February 28, 1984. Cited in *Report of the Presidential Commission on the Space Shuttle Challenger Accident* (Washington, D.C., 1986), vol. 1, p. 132.
[4] Thiokol Report, April 21, 1983. Cited in *Commission Report*, vol. 1, p. 133.

Challenger lifts off on the first night launch of the shuttle transportation system at 2:32 A.M., August 30, 1983. (NASA)

ated by the leak check history. Prior to January 1984, and STS 41-B, when the leak check pressure was 50 to 100 psi, only one field joint anomaly had been found during the first nine flights. However, when the leak check stabilization pressure was officially raised to 200 psi for STS 41-B, over half the shuttle missions experienced field joint O-ring blow by or erosion of some kind."[5]

[5] *Commission Report,* vol. 1, p. 134.

The commission added that the nozzle joint O-ring experience was similar. After the nozzle leak check pressure was increased form 50 to 100 psi, incidence of anomalies in the nozzle joint increased from 12 to 56 percent; and after nozzle leak check pressure was increased to 200 psi in April 1985, the incidence of erosion jumped to 88 percent.

On May 2, 1986, members of the commission questioned Russell and Marshall's Lawrence Mulloy about the putty blowhole predicament. Why did NASA and contractor engineers recommend increasing leak check pressure after it became obvious the change resulted in more erosion? The rationale for this was "corrective action," as Russell explained it, to make certain that putty was not (masking), a potential O-ring leak.

As an example, he cited the case of *Challenger* 51-B, wherein the left nozzle joint primary O-ring failed to seal and the secondary O-ring was eroded. It was after this event that leak check pressure at nozzle joints was raised from 100 to 200 psi to make sure that the primary O-ring would seal, following the procedure established for field joints a year earlier.

It was that event also that led Marshall to impose a launch constraint on all future shuttle flights, including *Challenger* 51-L. However, as mentioned earlier, the launch constraint was waived for each successive flight as it came to the launch pad. Commissioners asked why it was imposed at all.

Under questioning by Randy Kehrli, commission staff investigator, Russell was asked what he understood a launch constraint to mean.

Russell: My understanding of a launch constraint is that the launch cannot proceed without adequately— without everyone's agreement that the problem is under control.
Chairman Rogers: Under control meaning what? You just said a moment ago that you would expect some corrective action to be taken.
Russell: That is correct, and in this particular case on this 51-B nozzle O-ring erosion problem there had been some corrective action taken, and that was included in the presentation made as a special addendum to the next flight readiness review. And at the time we did agree to continue to launch, which apparently had lifted the launch constraint— would be my understanding.

"What was the corrective action taken?" asked commission member Dr. Arthur B. C. Walker, Jr., an astronomer and professor of applied physics at Stanford University.

Russell: One of the corrective actions was an increase in leak check pressure because one of the ways we thought that could happen—our analysis said that the blow-by across the O-ring which added to the erosion of that particular O-ring started right at ignition, indicating the O-ring didn't seal right from the very beginning.

And one of the ways that we thought that could have happened was that the putty could have masked our leak check, and we thought it was of the utmost importance to have a verified primary O-ring, and so we increased leak check pressure to 200 psi to make sure that we would blow through the putty, realizing that blowholes are not desirable either, but yet it is more important to know that you have a good O-ring and have some putty blow through than otherwise.

"Mr. Russell," asked commission member Dr. Eugene Covert, head of the Department of Aeronautics and Astronautics at MIT, "where did 200 psi come from? Is that twice 100?"

Russell: No. We did some bench testing of some putty and some gaps that showed. We tried at 50, 100, and 150 psi, and since we have a 10-minute stabilization period, the question was: can the putty hold these certain pressures for 10 minutes? In no case could the putty hold 150 psi for 10 minutes. And we put a factor on top of that of another 50 psi and that was the basis for that."
Dr. Walker: The analysis that some of our staff has done suggests that after you increase the test pressure to 200 pounds, the incidence of blow-by and erosion actually increased.
Russell: We realized that.
Walker: Did you realize that only after the accident?

At this point, Allan J. McDonald, manager of the solid rocket motor project at Thiokol, spoke up: "Could I try to answer some of that because I was very involved in this, and in fact this is the first time I realized that was a launch constraint. . . . That was the first time we had violated a primary O-ring."

McDonald went on to explain that the erosion that caused the violation on 51-B was different from anything seen previously. Blow-by at ignition had caused the erosion of the primary O-ring, he said, adding: "Therefore, we felt that we missed the leak check on that flight because the putty may well have masked it."

It was concluded that the circumstances of the seal failure on 51-B may have been unique, he said; still, a 200-psi leak check on nozzle as well as field joints would eliminate the potential for putty to mask the test, he said.

"That, I think was the basis for proceeding on with the next flight," he said. "We had to address it [erosion] prior to the next flight and I always felt that it could always become a launch constraint if it was not adequately understood and [if it was not] explained why this problem could not get worse on the next flight."

Marshall's booster manager, Lawrence Mulloy, described the reasons that had led Marshall management to impose the launch constraint on all future launches as the result of the 51-B booster seal findings. Testifying at the commission hearing on May 2, 1986, he said that the erosion of a secondary O-ring "was a new and significant event that we certainly did not understand." The evidence showed that "here was a case where the primary O-ring was violated and the secondary O-ring was eroded, and that was considered to be a more serious observation than any observed before."

Therefore, Mulloy and the Marshall Problem Assessment Committee had invoked the launch constraint to make sure a primary O-ring in a nozzle joint would not fail again. Members of the commission were not sure what it accomplished, inasmuch as it did not affect the launch schedule. According to 1980 Marshall documents, "all open problems coded criticality 1, 1R, 2 or 2R[6] . . . would be considered launch constraints until resolved."

Failure of the primary O-ring seal on the 51-B

nozzle joint was an open problem that Marshall sought to resolve by increasing the pressure of the leak check to make sure that the primary O-ring would seal. The 51-B nozzle joints had been leak-checked at 100 pounds per square inch pressure. Mulloy said it was concluded at Marshall that the putty could contain that pressure and would hide inability of the primary O-ring to seal if it was out of position.

Test data had shown that a leak check at 200 psi would always blow through the putty, although in doing so, it would leave holes through which hot engine gas could impinge on O-rings after ignition. But: "in always blowing through the putty, we were guaranteed that we had a primary O-ring that was capable of sealing," Mulloy said. "We already had that in the field joints at the time."

The putty predicament was a space-age version of Catch 22. To make sure that the primary O-ring would seal against leakage, Marshall's corrective action increased the incidence of erosion that threatened to cause leakage.

"Do you agree that the primary cause of erosion [in the O-rings] is the blowholes in the putty?" Dr. Walker asked Mulloy.

Mulloy : I believe it is, yes.
Dr. Walker: And so your leak check procedure created blowholes in the putty?
Mulloy: That is one cause of blowholes in the putty.
Dr. Walker: But in other words, your leak check procedure could indeed cause what was your primary problem. Didn't that concern you?
Mulloy: Yes, sir.[7]

Fixes

The rising incidence of seal damage in 1984 and 1985 persuaded headquarters to send two experts to Huntsville to find out what had gone wrong with the booster design after four years of operations. The in-

---

[6] *Commission Report*, vol. 1, p. 137. Whereas criticality 1 refers to a single-point failure causing loss of mission, vehicle, and crew, criticality 2 refers to failure leading merely to loss of mission.

[7] Testimony of May 2, 1986.

spectors were Irving Davids, an engineer, and William Hamby, deputy director of shuttle integration. They visited the Marshall Space Flight Center in July 1985 to discuss the O-ring problem with center engineers.

A memorandum written to L. Michael Weeks, deputy administrator for space flight (technical), by Davids indicates a range of opinions at Marshall on the cause and cure of the erosion problem. The prime suspect, Davids said, was the type of putty being used. He said that after the *Challenger* 41-B launch of February 3, 1984, the putty supplier went out of business. Putty obtained from a new source seemed to be more susceptible to environmental effects. Moisture made it "tackier."

But there was no consensus at Marshall, the report showed, on an effective fix. And the cause was ambiguous. It was Thiokol's position, Davids reported, that blowholes could be formed in the putty, allowing hot gas to reach the rings, in several ways. They could be formed during the assembly of motor segments at the Kennedy Space Center, during the leak check, or at ignition.

Thiokol, Davids reported, was considering several fix options. One was to remove the putty entirely. This option could be tested on a qualification (test) motor, referred to as "QM-5." Another option was to vary the putty lay-up to see if hot gas penetration would be prevented. Another was to substitute a putty filled with asbestos.

"Considering the fact there doesn't appear to be a validated resolution as to the effect of the putty," Davids said, "I would certainly question the wisdom of removing it [the putty] on QM-5."

Beyond this, Davids found that some Marshall engineers had reservations about the blowhole theory of erosion. His report said: "There are some Marshall Space Flight Center personnel who are not convinced that the holes in the putty are the source of the problem but feel that it may be a reverse effect in that the hot gases may be leaking through the seal and causing the hole track in the putty."

Davids noted five instances of primary field joint O-ring erosion and explained: "The erosion with the field joint primary O-ring is considered by some to be more critical than the nozzle joint due to the fact that during the pressure build up on the primary O-ring the unpressurized field joint secondary seal unseats due to joint rotation."

This problem, he said, "has been known for quite some time." In order to eliminate it, he went on, a capture feature was designed to clamp the tang with one leg of the clevis and prevent the secondary seal from lifting off. During discussions on this issue, he said, Marshall engineers were assigned to identify the timing of secondary O-ring unseating and the seating of the primary O-ring during the 600 milliseconds that joint rotation was taking place.

"How long it takes the secondary O-ring to lift off during rotation and when in the pressure cycle it lifts are key factors in the determination of its criticality," he said. "The present consensus is that if the primary O-ring seats during ignition and subsequently fails, the unseated secondary O-ring will not serve its intended purpose as a redundant seal. However, redundancy does exist during the ignition cycle, which is the most critical time."

The belief that secondary O-ring redundancy existed in the field joints during an early phase of the 600 milliseconds (0.6 seconds) of the ignition transient was an article of faith among management people at Marshall.[8] This belief supported an even broader view among some Thiokol and Marshall engineers that the field joint sealing system was redundant except in unusual circumstances, despite the criticality 1 reclassification that said officially that there was no redundancy. The commission report underlined the ambiguity of this situation: "Notwithstanding the view of some Marshall engineers that the secondary ring was not redundant, even at the time of the criticality revision, Marshall Solid Rocket Motor

[8] The ignition transient refers to the 600 milliseconds in which the propellant ignites and produces exhaust gas at increasing pressure until the pressure is stabilized.

program management appeared to believe the seal was redundant in all but exceptional cases." Not only Marshall management but also Thiokol held this view, the report added.[9]

This ambiguity may account for the rationalization among Marshall managers that it was safe to launch the shuttle despite persistent primary O-ring erosion. The thrust of the testimony the commission heard from Marshall executives was that the secondary O-ring would work if the primary failed during the ignition transient—at least most of the time.

According to the testimony and Marshall documents, tests had shown that there was a high probability of reliable secondary O-ring seating during the first 170 milliseconds of the ignition transient. That probability deteriorated from 170 to 330 milliseconds as motor pressurization produced joint rotation, opening up the O-ring extrusion gap and allowing the secondary O-ring to lift off metal mating surfaces. At 600 milliseconds, there was "a high probability of no secondary seal capability" in the field joints, according to the documentation.

In a commission hearing on May 2, 1986, Glynn Lunney, former shuttle program manager at the Johnson Space Center, described his view of the reclassification of the field joint seals to criticality 1 and of the waiver from headquarters that allowed the shuttle to continue flying. He said that he was involved in the approval of the waiver in March 1983.

"I was operating on the assumption that there really would be redundancy most of the time except when the secondary O-ring had a set of dimensional tolerances add up, and in that extreme case there would not be a secondary seal. So I was dealing with what I thought was a case where there were two seals unless the dimensional tolerances were such that there might only be one seal in certain cases."

At an earlier hearing on February 26, Dr. Judson A. Lovingood, deputy manager of the Marshall shuttle projects office, cited two conditions "you have to have before you don't have redundancy." He explained: "One of them is what I call a spatial condition which says that the dimensional tolerances have to be such that you get a bad stack-up, you don't have proper squeeze, etc. on the O-ring so that when you get joint rotation, you will lift the metal surfaces off the O-ring. All right, that's the one condition and that is a worst case condition involving dimensional tolerances.

"The other condition is a temporal condition which says that you have to be past a point of joint rotation, and of course that relates to what I just said, the joint has to rotate. So, first of all, if you don't have this bad stack-up, then you have full redundancy. Now secondly, if you do have the bad stack-up, you had redundancy during the ignition transient up to the 170 millisecond point or 300 millisecond point, whatever it is, but that is the way I understand the CIL" (critical items list).[10]

In the context of these views, as Davids had mentioned, the option for preventing the secondary O-ring from being unseated in a later phase of the ignition transient was the installation of the capture device to halt joint rotation. Although such a fix was widely discussed, it was not attempted.

The record shows uncertainty at headquarters and at the centers about what to do. A memorandum dated July 22, 1985, from Russ Bardos, a headquarters engineer, to David L. Winterhalter, chief of propulsion at headquarters, warned against a "quick fix" of the O-ring problem. Before any changes were started, he said, "we should make certain we fully understand the current design and problem."

The lack of a consensus among Marshall and Thiokol engineers about the causes and effects of seal erosion, about the seriousness of it in terms of flight safety, and about the actuality of redundancy resulted in uncertainty. These factors inhibited a sense of urgency to take prompt corrective action.

[9] *Commission Report*, vol. 1, pp. 126, 127.

[10] Transcript of the February 26, 1986 commission hearing, pp. 1700–1.

Bardos' memorandum seemed to express the predominant feeling at headquarters and at Marshall: "I do not believe the problem is so serious that the flight use of the SRM [solid rocket motor] should be prohibited," the memo stated.

The belief that the O-ring erosion problem could be worked through, one way or another, without interfering with the busy launch schedule kept the shuttle flying on spectacular missions.[11] O-ring erosion became rationalized as a persistent but not critical effect of the booster seal design.

More extensive O-ring damage was found after the launch of *Discovery* 51-C early in 1985. Primary O-ring erosion was found in the center segment field joint of the right booster, and soot had blown on the secondary O-ring. For first time, a heat effect was seen on the secondary O-ring. In the left booster, erosion and blow-by were seen on the primary O-ring of the forward segment joint. In addition, blow-by was seen on primary O-rings of the nozzle joints of both right and left boosters.

---

[11] These missions included *Challenger* 41-B in February 1984 when Bruce McCandless II and Robert L. Stewart took the first untethered space walks, demonstrated the MMU (manned maneuvering unit), and demonstrated the manipulator foot restraint on the orbiter's robot arm; *Challenger* 41-C in April 1984 when Dr. George D. Nelson and Dr. James D. Van Hoften captured, repaired, and redeployed the Solar Maximum Mission satellite that had sustained failure of critical scientific instruments since its launch in 1980; *Discovery* 41-D, August 30–September 5, 1984 when three satellites were deployed on a single mission for the first time and a 105-foot solar array was extended from the cargo bay, the largest structure ever raised from a spacecraft; *Challenger* 41-G in October 1984 when the technique of refueling a satellite in orbit was demonstrated for the first time; *Discovery* 51-A in November 1984 when two communications satellites were deployed and two others, Westar VI and Palapa B-2, which had been stranded in low orbit since February by upper-stage booster failures, were salvaged by Joseph Allen and Dale Gardner and returned to the factory for refurbishment and resale; *Discovery* 51-I, August 24–September 3, 1985 when Van Hoften and William F. Fisher retrieved, repaired, and redeployed the satellite Leasat 3, which had been stranded in low orbit when its upper-stage booster failed to ignite; *Atlantis* 61-B, November 26–December 3, 1985 when the first experiments testing the feasibility of erecting and assembling structures in space were carried out.

At this point in the flight experience, there was sufficient O-ring erosion experience to provide Marshall and Thiokol engineers with a "data base." From that base it was surmised that the erosion since 1981 could be accepted in flight operations. In their evaluation of erosion seen in 51-C, Thiokol engineers concluded that it was "consistent with the data," meaning that it had not become worse. Marshall agreed with that conclusion. It was recalled in the postflight analysis that similar erosion had been seen on 41-C and *Discovery* mission 51-A (launched November 8, 1984).

Out in the Cold

Erosion and blow-by of four primary O-rings and heat effect on a secondary O-ring in the 51-C boosters set a new damage record. The commission noticed that Thiokol made an extensive analysis of the problem for the flight readiness review of the next flight, 51-E.[12] Now for the first time, cold weather at launch time was considered as a factor in seal erosion and blow-by.

During the shuttle flight experience, the east coast of central Florida had been experiencing an unusual incidence of freezing weather in December and January. *Discovery* 51-C had been launched January 24, 1985 at 2:40 P.M. It was the coldest launch day in the program up to that time. At midafternoon, the temperature at the launch pad was 53° Fahrenheit, having risen after a cold night. The launch had been scrubbed on the previous day because of freezing conditions.

Thiokol's postlaunch analysis stated that "low temperature had enhanced the probability of blow-by." It added: "STS 51-C experienced the worst case temperature change in Florida history." It warned that the next flight could "exhibit the same behavior," but concluded that the "condition is acceptable."

However, at the flight readiness review for mission 51-E, the 18-page analysis was reduced to a one-page

---

[12] Combined with and flown as 51-D Apr. 12, 1985.

chart, the commission said. The report to Level 1 stated that the risk was acceptable "because of limited exposure and redundancy."

The notion that low temperature at launch contributed to seal erosion was based on the belief that it tended to increase the durometer (stiffness) of the rubber O-rings and decrease the springiness they required to seal joints.

Normally, although the primary O-ring was forced out of position by the leak check, it was forced back into sealing position immediately after ignition by gas pressure. On the other hand, the gas pressure that reseated the primary O-ring caused the joint rotation that unseated the secondary O-ring. Such was the scenario related with varying degrees of certitude by NASA and contractor witnesses. At any rate, the belief that the secondary O-rings in the field joints tended to become unseated was upheld by the reclassification of the field joint sealing system to criticality 1.

Irrespective of its actuality, this scenario set the stage for the visualization of the effect of cold weather. If it was stiffened by cold, the primary O-ring would move more slowly to seal the joints under engine gas pressure, thus allowing more erosion and blow-by. If cold and stiff enough, the primary O-ring might fail to seal.

A pertinent illustration of the effect of cold weather on the performance of the O-ring seals was depicted by the seal damage found in the 51-C boosters. Its significance was described to the commission at its hearing on February 25, 1986 by Roger M. Boisjoly, senior scientist on Thiokol's seal task force. He said:

SRM 15 [on 51-C][13] actually increased [our] concern because that was the first time we had actually penetrated a primary O-ring on a field joint with hot gas, and we had a witness of that event because the grease between the O-rings was blackened just like coal. . . .

And that was so much more significant than had ever been seen before on any blow-by on any joints.

Earlier, he explained that blow-by erosion occurs when the primary O-ring is attacked by hot gas at the beginning of the ignition transient cycle "and it is eroding at the same time it is trying to seal." These phenomena were "a race between—will it erode more than the time allowed to have to seal."[14]

"The fact was," Boisjoly added, "that now you introduced another phenomenon. You have impingement erosion and bypass erosion, and the O-ring material gets removed from the cross-section of the O-ring much faster than when you have blow-by."[15]

Boisjoly and his colleagues at Thiokol equated the exceptional 51-C damage with the relatively low temperature of launch day. The experience of 51-C was a turning point in the performance history of the booster seals. From then on, the frequency of seal damage increased. And from then on, some of the Thiokol engineers regarded the temperature at which 51-C was launched as the lowest at which the shuttle could be launched safely.

Marshall's continuing review of joint problems concluded that although the primary O-ring in the right booster's center field joint had been damaged enough to expose the secondary O-ring to hot gases, the secondary O-ring had not been eroded enough to affect its sealing capability. The review stated:

Primary O-ring erosion is expected to continue, since no corrective action has been established that will prevent hot gases from reaching the primary O-ring cavity. Steps have been taken to assure that the secondary O-ring will be seated, and analyses have indicated that under a worst case situation, erosion of the second-

---

[13] Thiokol's solid rocket motor no. 15 in the shuttle series.

[14] Testimony before the commission on February 14, 1986, p. 1202.

[15] Impingement erosion occurs when a focused hot gas jet strikes the surface of an O-ring that has sealed and burns part of it away. Blow-by erosion occurs before the O-ring has sealed the joint gap, when hot gas flows around the edge of the ring and erodes it.

ary O-ring will not be severe enough to allow a leak path past the secondary O-ring.

Thiokol conducted O-ring resiliency tests early in 1985, and a report on August 9 concluded that O-ring resiliency "is a function of temperature and rate of case expansion" (presumably the ballooning of the metal case under gas pressure of 1,000 psi that caused the joint to open). At 100° F, the report said, the O-ring maintained contact with the metal. At 75° F, the O-ring lost contact for 2.4 seconds (as the case expanded). At 50° F, the O-ring did not reestablish contact in 10 minutes. At that time, the test was terminated.

The Presidential Commission made a brief study of thermal O-ring distress[16] in relation to temperature. It reported that only three instances of thermal distress appeared after 20 launches at O-ring temperatures of 66° F or higher. All four flights at O-ring temperatures of 63° F or lower showed thermal distress.

The commission concluded: "Consideration of the entire launch temperature history indicates that the probability of O-ring distress is increased to almost a certainty if the temperature of the joint is less than 65."

Closing Out

By December 1985, O-ring erosion and blow-by had become chronic. Of the 10 shuttle flights after 51-C, 7 had O-ring erosion by blow-by or both, and the eighth, *Challenger* 51-L, was destroyed. Only two flights appeared to have escaped seal damage. These were 51-F, a *Challenger* mission launched July 29, 1985, and 51-J, an *Atlantis* flight launched October 3, 1985. Both launches took place with the highest joint temperatures of the year, 51-F at 81° F and 51-J at 79° F.

Despite the apparent correlation of ambient temperature and seal damage on launch day, low temperature was never considered as a launch constraint by

[16] Evidence of O-ring burns or discoloration from hot gas.

NASA. Late in December, as launches continued in colder winter weather than is expected in Florida, the unresolved O-ring problems that both Marshall and Thiokol had been continuously tracking were suddenly closed out at Marshall as if they had been solved. Memoranda showed that the action was requested by Thiokol, but testimony indicated that it was prompted at Marshall.

In the light of the *Challenger* accident, closing out the O-ring problem seemed imcomprehensible. The commission sought to determine why it was done. The records showed that Brian Russell of Thiokol sent a memorandum to Allan J. McDonald, Thiokol's solid rocket motor project manager, on December 6, 1985 requesting closure of the O-ring erosion problems. Russell gave 17 reasons for closing them out. These included test results, future test plans, and efforts of a special Thiokol task force to resolve the O-ring problems, the commission report said.

Four days later, the report continued, McDonald wrote a memorandum to Lawrence O. Wear, manager, solid rocket motor project, at Marshall asking for closure of the O-ring problem. McDonald's memo noted that the O-ring problem would not be fully resolved for some time, the commission report said, but he enclosed a copy of plan Thiokol had developed for improving the seals.

At the commission hearing on May 2, 1986, Russell was asked what he meant by the memorandum. He replied there was a request to close out problems that had been open for a long time. As he recalled, he said, it came from the director of engineering.

Dr. Walker: That was the director of engineering at the Marshall Space Flight Center?

Russell: Yes, at Marshall Space Flight Center . . . my recollection was that Mr. Kingsbury would like to see these problems closed out.[17] Now, the normal

[17] James E. Kingsbury, director of science and engineering, solid rocket motor, office of the associate director for engineering, Marshall Space Flight Center.

method of closing them out is to implement the corrective action, verify the corrective action, and then problem is closed; it comes off the board and is no longer under active review.

Chairman Rogers asked what was being done to fix the problem. Russell said that a task force of full-time people had been set up at Thiokol, and had made some engineering tests. Russell was a member of that team. He said: "We were trying to develop concepts. We had developed some concepts to block the flow of hot gas against the O-ring to the point where the O-ring would no longer be damaged in a new configuration. And we had run some cold gas tests and some hot gas motor firing tests and were working toward a solution of the problem, and we had some meetings scheduled with the Marshall Space Flight Center—"

Chairman Rogers: Can I interrupt? So you're trying to figure out how to fix it, right? And you're doing some things to try to help you figure out how to fix it. Now, why at that point would you close it out?
Russell: Because I was asked to do it.
Rogers: I see. Well, that explains it.

"It explains it, but really doesn't make any sense," said Commissioner Robert W. Rummel, aerospace engineer and former vice president of Trans World Airlines. "On one hand you close out items that you've been reviewing flight by flight that have obviously critical implications, on the basis that after you close it out, you're going to continue to try to fix it. So I think that what you're really saying is you're closing it out because you don't want to be bothered. Somebody doesn't want to be bothered with flight-by-flight reviews, but you're going to continue to work on it after it's closed out."

On January 23, 1986, five days before the *Challenger* disaster, an entry was made in the Marshall Problem Reports that the "problem is considered closed." When Chairman Rogers asked Lawrence Mulloy about that, the Marshall solid rocket booster manager asserted that the problem was closed out in error. "The

people who run this problem assessment system [at Marshall] erroneously entered a closure for the problem on the basis of this submittal from Thiokol," Mulloy said.

Thus, the O-ring problem did not come up as an open launch constraint in the *Challenger* 51-L flight readiness review, Mulloy said. However, as the records showed each time the launch constraint was invoked after the 51-B mission, it was waived up to the December problem close-out.

Commissioner Eugene Covert asked Mulloy: "at the time this joint was conceived of, did you envision then that the O-rings would be eroded to this extent?"

Mulloy: No.
Covert: So, in some way, the acceptance of this erosion as a fact of life represented a departure from margins of safety that you originally had in mind when you were designing it?
Mulloy: Yes, sir. It was treated as an anomaly.

The Search for Solutions

Following the August 19, 1985 briefing on seal problems to NASA headquarters by Thiokol and Marshall program managers, Robert Lund, vice president of engineering at Thiokol, announced the formation of a Thiokol O-Ring Task Force. Its function was to investigate O-ring problems and recommend short- and long-term solutions.

Thiokol's supervisor of structures design, A. R. Thompson, recommended near-term fixes. One was to use thicker shims in securing tang and clevis fit and to increase O-ring thickness. In a memorandum to a colleague, he stated that several long-term solutions looked promising but would take several years to implement. "The simple short-term measures should be taken to reduce flight risks," he said.[18]

---

[18] Memorandum to S. R. Stein, project engineer, of August 22, 1985.

Potential Long Term Solutions--Field Joints

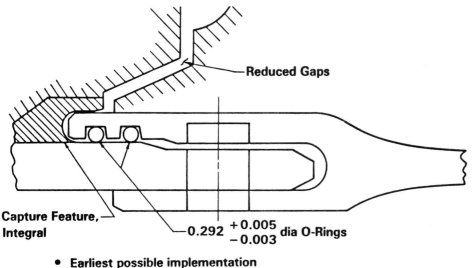

- **Earliest possible implementation**
  - **Capture feature - 27 months, STS-81N (SRM-67)**
  - **Reduced gaps - 12 months, STS-71P (SRM-52)**

One of the numerous field joint design changes proposed by Morton Thiokol engineers to NASA at a headquarters conference and briefing in August 1985. This one showed a capture feature that would keep the gap between the inner leg of the clevis from opening during the ignition transient when hot gas pressure inflated the joint before the pressure was stabilized. Also proposed was a reduction in the insulation gap (filled with putty in the operating design) and an increase in the diameter of the O-rings from 0.28 to 0.292 inches. Implementing these changes would take 12 to 27 months.

- **Potential near term conditions - field joint**

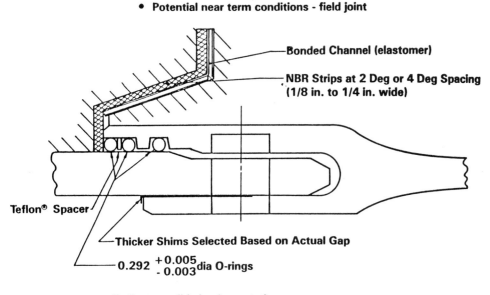

- **Earliest possible implementation**

  - **Clevis modification 12 months, STS-71J (SRM-46)**

Another Thiokol proposal called for three 0.292-inch O-rings, thicker shims to reduce the gap between tang and clevis, and replacement of the putty with nitrile strips and a bonded channel elastomer. Modifying the clevis would take a year.

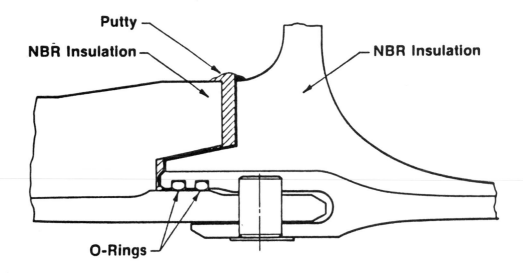

Putty

NBR Insulation

NBR Insulation

O-Rings

MORTON THIOKOL, INC.
Wasatch Division

INFORMATION ON THIS PAGE WAS PREPARED TO SUPPORT AN ORAL PRESENTATION
AND CANNOT BE CONSIDERED COMPLETE WITHOUT THE ORAL DISCUSSION

The existing joint. Putty was used to fill the gap where the segments were joined and to protect the O-rings from being burned or eroded by hot gas. However, hot gas occasionally reached the O-rings through holes punched through the putty by high-pressure leak checks. The NBR insulation is nitrile butadiene rubber and forms a buffer between the propellant (PBAN) and the steel motor case.

By the end of August, Thiokol's seal task force compiled a list of 43 design concepts to solve the seal problem and made plans to fire a static test motor with several O-ring configurations, the commission report said. Other Thiokol documents perused by the commission showed that the seal problem task force was having problems. In September, complaints surfaced about administrative delays and lack of cooperation.

One complaint from Boisjoly dated October 4, 1985 warned that even "NASA perceives that the team is blocked in its engineering efforts to accomplish its task."

R. V. Ebeling, manager of Thiokol's solid rocket motor ignition system, sent a memorandum to McDonald beginning with the plea, "HELP!" He complained that the task force was "constantly being delayed by every possible means." It was limited to a group of 8 to 10 engineers, some encumbered with other significant work, he said. Marshall was correct, he said, in stating that "we do not know how to run a development program."[19]

Professor Covert asked Ebeling at the May 2 hearing: "did you have an increasing concern as you saw the tendency first to accept thermal distress and then to say, well, we can model this reasonably and we can accept a little bit of erosion, and then, etc., etc.? Did this cause a feeling of, if not distress, then betrayal in terms of your feeling about O-rings?"

Ebeling: I'm sorry you asked that question.
Covert: I'm sorry I had to.
Ebeling: To answer your question, yes. In fact I have been an advocate—I used to sit in on the O-ring task force and was involved in the seals, since Brian Russell worked directly for me, and I had a certain allegiance to this type of thing anyway, that I felt that we shouldn't ship any more rocket motors until we got it fixed.

[19] Memo, Ebeling to McDonald, October 1, 1985, *Commission Report.*

Kennedy Space Center inspectors examine two segments of the left-hand solid rocket booster that had been assembled for shuttle mission 61-G, the mission scheduled after 51-L. The boosters were disassembled in April 1986 and inspected for flaws in the assembly process. (NASA)

Covert: Did you voice this concern?
Ebeling: Unfortunately, not to the right people.

The De-Stacking Experiment

The flight that was to follow 51-L was 61-G, an *Atlantis* mission, for which the solid rocket boosters had been mated in the Kennedy Space Center's Vehicle Assembly Building by the time of the accident. Could improper stacking of the booster segments have been a contributing factor to the *Challenger* accident? In order to determine that, the acting NASA administrator, William Graham, ordered the Data Design and Analysis Task Force to take the boosters apart and look for errors in stacking the segments, such as out-of-roundness. The

This photo shows how booster segments are stacked on top of the mobile launcher platform in the Vehicle Assembly Building. (NASA)

disassembly inspection focused on the left booster aft field joint and all three field joints on the right booster.

On May 1, 1986, the task force reported to Graham that metal shavings and other debris had been found in the pinholes where the segments were fastened together; that water was found in the bottom of a clevis in one joint; that blowholes were found in the putty; and that the putty had not adhered uniformly to the upper sur-

face. No "distress" was found in the O-rings nor tang and clevis parts of the joints.[20]

As mentioned earlier, putty blowholes had been determined as the pathways through which hot gases could impinge on the O-rings and blow soot by them. Water in the U-shaped clevis could freeze in cold

[20] Historical Summary, Mission 51-L Mishap Investigation, p. 3–89.

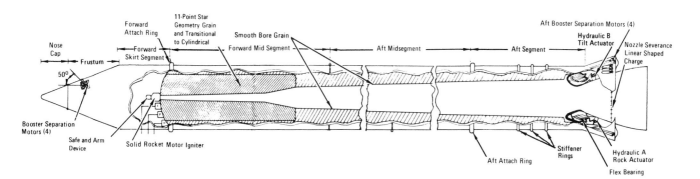

The motor, 12.17 feet in diameter, is the largest solid rocket motor ever built. This view shows the propellant with 11-point star-shaped perforation in the forward motor and the double truncated-cone shaped perforation in the aft segments. This arrangement was designed to provide high thrust at ignition and then reduce the thrust by one-third about 50 seconds after liftoff to avoid overstressing the shuttle as it passes the region of maximum dynamic pressure (Max Q). (Rockwell)

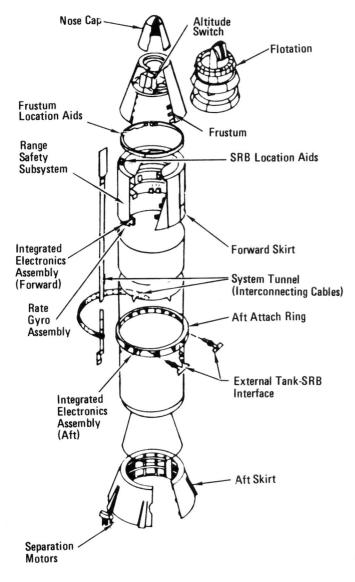

Other major components of the booster. (Rockwell)

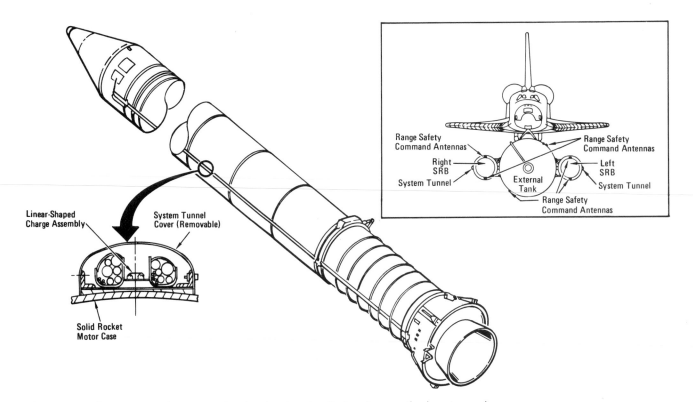

The Range Safety Destruct System on the shuttle, showing its location on the booster and the location of antennas that receive the "destruct" radio signal from the range safety officer. (Rockwell)

weather and interfere with O-ring seating, a possibility that had been speculated upon as a factor in the *Challenger* accident. However, no official comment about that was made in the de-stacking report.

Finding

The commission concluded that the *Challenger* accident originated with decisions made in the design of the joint and in the failure by both Thiokol and NASA's solid rocket booster project offices to understand or respond to facts obtained during testing.

Neither Thiokol nor NASA responded adequately to internal warnings about the faulty seal design, the commission said, adding:

Furthermore, Thiokol and NASA did not make a timely attempt to develop and verify a new seal after the initial design was shown to be deficient. Neither

organization developed a solution to the unexpected occurrences of O-ring erosion and blow-by even though this problem was experienced frequently during the shuttle flight history. Instead, Thiokol and NASA management came to accept erosion and blow-by as unavoidable and an acceptable flight risk.

Specifically, the commission found that:

1. The joint test and certification program was inadequate. There was no requirement to configure the qualification test motor as it would be in flight, and motors were static-tested in a horizontal position, not in the vertical flight position.

2. Prior to the accident, neither NASA nor Thiokol fully understood the mechanism by which the joint sealing action took place.

3. NASA and Thiokol accepted escalating risk apparently because "they got away with it last time." As Commissioner Richard P. Feynman, the Nobel laureate physicist, observed, decision-making was

The aft segment of the solid rocket booster and its attachment to the external propellant tank are shown in this work photo in the Vehicle Assembly Building. The scaffolding to the left of the booster supports heat shield repair on the orbiter wing. (NASA)

a kind of Russian roulette . . . [in which the shuttle flies with O-ring erosion] and nothing happens. Then it is suggested, therefore, that the risk is no longer so high for the next flights. We can lower our standards a little bit because we got away with it last time. . . . You got away with it, but it shouldn't be done over and over again like that.

4. NASA's system for tracking anomalies for flight readiness reviews failed in that, despite a history of persistent O-ring erosion and blow-by, flight was still permitted. It failed again in the strange sequence of six consecutive launch constraint waivers prior to 51-L permitting the shuttle to fly without any record of a

Two segments of the left-hand solid rocket booster originally assembled for mission 61-G are disassembled from the booster stack preparatory for inspection to determine whether there were flaws in the assembly process of 51-L. (NASA)

waiver or even of an explicit constraint. Tracking and continuing only anomalies that are "outside the data base" of prior flight allowed major problems to be removed from and lost by the reporting system.

5. The O-ring erosion history presented to Level 1 at NASA headquarters in August 1985 was sufficiently detailed to require corrective action before the next flight.

6. A careful analysis of the flight history of O-ring performance would have revealed the correlation of O-ring damage and low temperature. Neither NASA nor Thiokol carried out such an analysis; consequently they were unprepared to properly evaluate the risks of launching the 51-L mission in conditions more extreme than they had encountered before.

5. The Flaw

The Presidential Commission on the Space Shuttle *Challenger* Accident began public hearings on Thursday morning, February 6, 1986 in the auditorium of the National Academy of Sciences, Washington, D.C., William P. Rogers, chairman, presiding. The first witness was Jesse Moore, associate administrator for space flight. He chaired the 51-L Data and Design Analysis Task Force that had been organized only the day before.

At the outset of the hearing, Rogers made it clear to the audience, principally news media representatives, that the commission was now in full charge of the *Challenger* accident investigation. It followed that NASA's role was simply to gather facts.

Rogers referred to the executive order that created the commission "because we want to stick very closely to the instructions that we received from the President." These were to investigate the accident, review the circumstances surrounding it to establish the probable cause or causes, and develop recommendations for corrective or other action based on the commission's findings and determinations. The commission was directed to submit its final report to the President and the administrator of NASA within 120 days of the order. That would be June 6, 1986.

Moore related that *Challenger* lifted off pad B at 11:38 A.M. after a two-hour delay. He did not have any concerns about the temperature other than those about ice on the pad structure, he said.

"The actual flight, the ascent appeared normal," he said, "based on our initial quick look for the first 73 seconds, and it went through its main program roll maneuver where the shuttle rolls from its initial launch configuration; through its maximum dynamic pressure . . . and then the throttle down and throttle back up of the shuttle main engines.

"The vehicle again appeared to be performing nominally at 104 percent thrust, at approximately 1,200 miles an hour at approximately 47,600 feet when all our telemetry stopped, and at that point in time we observed the breakup from the ground."

The commission heard testimony also from Acting Administrator Graham and Arnold D. Aldrich, manager of the National Space Transportation System Program at the Johnson Space Center, Houston. Graham, Moore, and Aldrich represented the command levels of space shuttle management and were in control of the launch. Except for citing the puff of smoke displayed on launch film as a possible clue to failure in the right booster, none of them indicated a probable cause. The puff of smoke was the only lead.

Moore displayed Vu-Graphs that appeared to show the plume coming from the right booster at 59.2 seconds and spreading until 73 seconds when the external tank appeared to blow up. Until more detailed study of the ascent film and greater enhancement of it were done, he said, the precise location of the plume could not be fixed.

"Which segment of the solid rocket would that be?" asked Robert Hotz, the retired editor of *Aviation Week and Space Technology*.

"We don't know," Moore replied. He indicated an area where the aft and the aft-center segments were

The Rogers commission hears testimony at an open hearing on March 7, 1986 at the Galaxy Theater of the Visitors Information Center, an annex of the Kennedy Space Center. Commissioners are *(left to right)*, front row: David C. Acheson, Richard P. Feynman, Neil A. Armstrong, Chairman William P. Rogers, Sally K. Ride, Alton G. Keel, Jr., commission executive director; back row: Joseph Sutter, Robert W. Rummel, Major General Donald J. Kutyna, Eugene E. Covert, and Arthur B. C. Walker, Jr. Witnesses on the left are *(left to right)* Thomas Moser, director of engineering, Johnson Space Center; J. Wayne Littles, associate director of engineering, and Jack Lee, deputy director, Marshall Space Flight Center. (NASA)

joined and where the structural attachment of the booster and the external tank were located. "We don't know whether it's the aft-center segment. We don't know for sure it is the SRB."

The Feynman Experiment

Most of that first session was devoted to a description of the shuttle and its main components: the orbiter, the boosters, and the external tank. No one speculated on the significance of the puff of smoke, although the news media reported film evidence that it signified a leak of hot booster gas that pierced the aluminum skin of the external propellant tank and set it ablaze.

The commission reconvened in public session on February 11, when the effect of the unusually cold weather at launch day on the synthetic rubber (Viton) O-ring seals was discussed. A lucid description of the solid rocket boosters was provided by Lawrence B. Mulloy, the SRB project manager at the Marshall Space Flight Center. A veteran engineer, Mulloy had managed the program since November 1982.

Mulloy explained that the solid rocket motor consisted of four casting segments each 24 feet long and 146 inches in diameter. The casting segment itself was made up of two 12-foot segments that were joined at the factory. Because insulation is applied after the joint is made, there is no discontinuity in it at the factory joint.

"On the field joint, however," he explained, "there is this discontinuity in the insulation, since you have to put it together at KSC, at the Kennedy Space Center. And this gap between the insulation is filled with a zinc chromate abestos-filled putty. That putty is laid up in strips prior to assembly.

"We use strips of putty an eighth-inch and quarter-inch thick and an inch to an inch and a half wide. . . . you have a good fill of the putty between the insulation surfaces, but . . . it does not extrude down into the O-ring gap such that it would tend to unseat the O-ring."

The O-rings, he said, were Viton rubber, made by Parker O Seal, Culver City, California. They were 280 thousandths of an inch (0.28 inches) in diameter. There were two O-rings at each joint. The segments were assembled with the clevis end up and the tang end down. Then the joint was fastened with 177 steel pins.

Referring to his chart, Mulloy called attention to a clip on the outboard leg of the clevis. It was a 32 to 36 thousandths of an inch shim. Its purpose was to "maximize the O-ring compression or the squeeze on these O-rings" between the inboard leg of the clevis and the tang. Mulloy then described how the O-rings functioned and the aspect of their resiliency in sealing the joints.

On that subject, commission member Richard P. Feynman, professor of theoretical physics at the California Institute of Technology, a 1965 Nobel laureate in physics, offered a comment for Mulloy.

"We spoke this morning," he said, "about the resiliency of the seal and if the material weren't resilient, it wouldn't work in the appropriate mode but would be less satisfactory, in fact it might not work well. I took this stuff that I got out of your seals and I put it in ice water and I discovered that when you put pressure on it for a while and then undo it, it doesn't stretch back. It stays the same dimension. In other words, for a few seconds at least—and more seconds than that—there is

no resilience in this particular material when it is at a temperature of 32° [Fahrenheit]. I believe that has some significance for our problem."

Chairman Rogers said that he thought it was an important point. The simplicity of the experiment in contrast to the complexity of seal dynamics aroused a titter of amusement in the audience. But the experiment was no joke. It was a teaching demonstration that illuminated the core of the probable cause of the accident if, indeed, the seals had failed as suspected.

Dr. Feynman admitted that "this is not the way to do such experiments," but it indicated that "the stuff looked as if it was less resilient at lower temperatures, in ice." He asked Mulloy: "Does your data agree with this feature, that the immediate resilience, that is, within the first few seconds, is very, very much reduced when the temperature is reduced?"

"Yes, sir," said Mulloy. "In a qualitative sense. I just can't quantify that at this time."

Feynman: . . . was there some kind of a temperature limit, then, on when you would say a seal was safe? Was there some kind of criterion that if the seal was lower than a certain temperature we cannot consider it safe enough?

Mulloy: Yes, sir. The data that we had—as I stated, we were running extensive testing to understand the response of the seal under the specific conditions of 51-L, but the data available to us was the procurement specification for the material that says that it will operate from minus 30 to 500° F. And also, a great deal of test data in motors down at temperatures where you can see a difference in the resilience of the seal, for instance going from 75° down to 50°, you can see that. We had these data available.

Feynman: I just wanted to know, before the event, from information that was available . . . was it fully appreciated everywhere that this seal would become unsatisfactory at some temperature and was there some sort of a suggestion of a temperature at which the SRB shouldn't be run?

Mulloy: Yes, sir. There was a suggestion of that, to

Kennedy Space Center Director Richard G. Smith *(right)* briefs members of the Rogers commission on shuttle assembly in the Vehicle Assembly Building during their visit to the Kennedy Space Center, February 14, 1986. To his left, clockwise, are Sally K. Ride, Chairman William P. Rogers, Alton G. Keel, Jr., commission executive director, and Robert W. Rummel. (NASA)

answer the first question. First, the data that was presented, it was the judgment that under the conditions that we would see on launch day, given the configuration that we were in, that the seal would function at that temperature. That was the final judgment.

The Cook Memoranda

The star witness of the day was Richard C. Cook, NASA budget analyst. It was the publication of his memos revealing concern among headquarters engineers about the booster seals that had focused public

KSC Director Richard G. Smith *(right)* shows Commission Chairman William P. Rogers how heat shield tiles are being replaced on the orbiter *Discovery* during the commission's tour of the space center, February 14, 1986. (NASA)

attention on the probability of seal failure as the cause of the accident.

"Mr. Cook" Rogers began, "the commission asked you to appear today because of recent stories concerning particularly a memorandum which you wrote on the 23d of July, 1985. . . .

Cook related his background as a policy analyst for the federal government since 1970. He had worked at the Civil Service Commission and the Food and Drug Administration; then at the White House Consumer Affairs Council in both the Carter and Reagan administrations. He left government for a while to work on a defense intelligence hardware project at TRW, Inc. and was engaged by NASA as a resource analyst in the

Members of the Rogers commission examine a propellant-loaded solid rocket booster segment in the Vehicle Assembly Building during their tour of the Kennedy Space Center on February 14, 1986. *Left to right:* Arthur B. C. Walker, Eugene E. Covert, Robert B. Hotz, Chairman William P. Rogers, Sally K. Ride, and Alton G. Keel, Jr. (NASA)

comptroller's office. In this capacity, he said, he was asked to look into potential budget "threats" arising out of problems with hardware that might have to be replaced or redesigned in the shuttle.

In talks with engineers at the Kennedy and Marshall centers, he said, it became apparent to him that there were concerns about the O-ring seals in the solid rocket boosters. These had safety as well as budget implications. Every month, an O-ring erosion or charring problem was on a list of budget threats, sometimes first on the list, he said.

"There was no question that the O-ring problem was considered a potential budget threat month after month from the summer [1985] and on into the fall," he said. ". . . as I understand it . . . it was, even when it went up to the administrator, it was also listed on that presentation as a budget threat."

He said it had been mentioned that an effort was going to be made to keep the secondary O-ring from unseating in flight as a result of joint rotation after motor ignition. "And so there was a lot of concern about how we could get redundancy back in that joint without having to throw away half-million-dollar SRB segments and start all over again with redesigning and recasing them. There was a 13-month lead time if you wanted to order a new segment from the manufacturer, and so if you had to throw this stuff out you had a problem."

Cook said that he was aware of reports of erosion on O-rings. While he was not competent to comment on Mulloy's presentation (on seals), he said, from his perspective "as the guy who is supposed to be watching this issue for the comptroller's office" it was his understanding that there was at least some erosion going on in 1985 in the O-rings.

Rogers asked Cook if the focus of his attention was primarily budgetary. It was, Cook replied.

"And to summarize it," said Rogers, "you were, I gather, thinking about whether, if changes were required

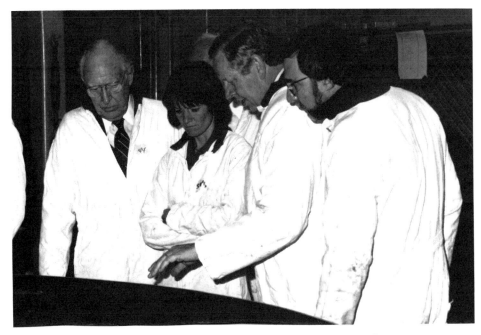

During the commission's extended visit to the Kennedy Space Center, February 13–14, 1986, members inspected booster segments. *Left to right* are Commission Chairman William P. Rogers, Astronaut Sally Ride, Kennedy Space Center Director Richard G. Smith, and Alton G. Keel, commission executive director. (NASA)

In the foreground of commissioners touring the Vehicle Assembly Building at the Kennedy Space Center February 14, are Commission Chairman William P. Rogers, Vice Chairman Neil Armstrong, and Astronaut Robert L. Crippen, member of the 51-L Data and Design Analysis Task Force. (NASA)

Members of the Presidential (Rogers) Commission inspect a solid rocket booster segment loaded with propellant in the storage chamber adjacent to the Vehicle Assembly Building during their tour on February 14. (NASA)

for safety reasons or any other reasons, you had to think about how much it would cost.''

Cook: Yes.

Rogers: And therefore your questioning of people in NASA was in connection with that budgetary matter?

Cook: Exactly.

Rogers: You didn't, I assume, make any attept to weigh budgetary considerations and safety considerations, did you?

Cook: Not at all.

Rogers: Well, since the accident occurred have you had discussions with people about your memorandum of July 23, 1985?

Cook: Yes. Particularly since it showed up in the newspapers.

Rogers: Did you have discussions with people before it showed up in the newspaper?

Cook: I had given it to my boss just as a matter of giving him documentation. [He referred to Michael B. Mann, chief, Shuttle Transportation System Resources Analysis Branch, Office of the Comptroller.]

Rogers: But to no one else?

Cook: Someone else? Well, my boss and the other former SRB analyst that I worked with very closely.

Rogers: Anyone out of the office?

Cook: No.

Rogers: And so you were not involved in the publication of the documents yourself?

Cook: No, I was not.

The chairman continued with this prosecutor-style interrogation to probe a second memorandum about the accident from Cook.

Roger: What prompted that?

Cook: The heat of the moment. . . .

The second memorandum was written by Cook a week after the accident and dated February 3, 1986. It stated at the outset: ''There is a growing consensus that the cause of the Challenger explosion was a burn-through in a solid rocket booster at or near a field joint.''

Cook explained that although the evidence pointing to the SRB as the cause of the accident was either circumstantial or based on photographs, the O-ring problem had to be taken care of before ''we could look at the shuttle program as being completely resolved.'' Other budgetary issues came up in that connection, he said.

Rogers: You still were doing it in terms of the budget? I mean, was that the purpose of writing the memorandum after the accident?

Cook: Yes.

Rogers: Did you have reason to think that efforts would not be made after the accident to investigate it thoroughly?

Cook: No. In fact, I knew that across the street [at NASA headquarters] they were doing the same analyses we were doing.

Rogers: Then will you explain again why you wrote the memorandum?

Cook: To document what I felt were all of the budgetary implications of the situation.

Rogers: Only budgetary? Was there any other purpose? Well, if you'd rather not answer, that's all right. I'm just curious about why you wrote the memorandum. I mean, it doesn't sound as if you had budgetary considerations in mind. It sounds differently. But I wanted you to have an opportunity to tell the commission why you wrote it.

Cook: I wrote it because I felt that it was a serious enough situation that I didn't think that until these various issues were resolved with the SRBs that I had been involved with—that they had to be taken care of before the shuttle could safely continue.

Rogers: Did you think your engineering experience based upon the short time you had been with NASA improved your ability to pass judgment on what others had decided? Well, here again, I don't really want to press you. Do you have anything else to tell the commission?

After lunch, Cook returned to the witness stand. Rogers said that he had talked with Cook during lunch and had learned that the budget analyst had been asked to prepare the postaccident memorandum of February 23. Nevertheless, Rogers said, Cook had testified that he wrote it in the heat of the moment. It seemed to observers that the chairman had gained an initial idea that Cook had written the second memorandum on his own, as if blowing the whistle. If so, the impression was dismissed during lunch, and Rogers invited Cook to tell the commission how he reacted to the engineers at NASA.

Cook testified it was a job requirement to try to understand as much as possible about the engineering side. The budget analyst played a middleman role, trying to translate engineering into language that could be used for cost and price analysis in budget presentations to Congress. He said his office was required to come up with fast estimates of the cost of suspected hardware failures. The boosters and external tank feed into his area of analysis, he said, and he tried to pull together all the budget implications he could think of.

One of them was the capture feature on the boosters that would prevent the joint from rotating during the ignition transient and would keep the secondary O-ring in position to seal. This improvement was to be tested on Qualification Motor 5. Another item was field joint putty. The qualification of a new putty (that was being considered) would be a major unbudgeted cost item.

"Any effects of environment and weather factors on putty and O-rings have design implications which require further investigation," he said. Eight SRB segments were in the manufacturing process, and any action to retrofit them with a capture feature or other new design feature had to be taken into account.

"If the capture feature cannot be qualified for the steel segments and we have to re-engineer the field joints—there are 6 per flight set that one way or another will have to be re-engineered—if that is the case, the lead time to tell the manufacturer to make a new segment . . . is about 13 months.

"So, based upon all of those considerations, if we went through with this program I was projecting—and again this is just a budgetary analyst estimate—I was projecting a nine-month or longer suspension of flights to deal with these things."

The hard thrust of Rogers' earlier interrogation of Cook softened after the chairman realized that the analyst was acting under orders. He told Cook:

"Now I asked you this morning about the reason

for the [second] memorandum [of] February 3. And it subsequently turns out that you were asked to provide a memorandum to the effect along the lines that you have reached in your summary just now. In other words, you were asked to do that by Mr. Mann?"

Cook: Yes, sir.

Rogers: And the memorandum that is dated February 3 that you have just summarized was in response to that request, and I think your summary of the memorandum is very good and it explains to the commission your motivation.

Rogers had one more question, about a portion of the memo saying: "It is also my opinion that the Marshall Space Flight Center has not been adequately responsive to headquarters' concerns about flight safety, that the Office of Space Flight has not given enough time and attention to the problems with SRB safety raised by senior engineers in the Propulsion Division. And those engineers have been improperly excluded from the investigation of the Challenger disaster."

Rogers asked: "In the light of the work of this commission and the investigations that are being conducted now at Kennedy, are you still of that view?"

Cook admitted that he had editorialized but said he was not prepared to go into details.

"I think the thing that concerns me most," said Rogers, "is whether you have confidence now that the investigations are being properly conducted."

Cook said that he would be more specific in answering this if he had access to his files and time to write. "But let me say this. The last item—frankly, I was amazed that when this incident occurred, the engineers in Washington were over there in their offices getting data on the investigation from the newspapers and now and then from phone calls from guys down at Kennedy about what was being found. These were the top propulsion engineers who prepared reports for the Office of Space Flight and for the administrator and for us. I just couldn't understand why that group wasn't down

there going through the data and looking at the photos and everything else."

Rogers then said that he understood that Cook was changing jobs and leaving NASA. Cook acknowledged that he had had an offer from the Treasury Department where he had worked before. He said he would report to work at Treasury the following week. "It doesn't have anything to do with this," Cook said.

Associate Administrator Jesse Moore was asked to comment on Cook's memoranda. He said that the July 23, 1985 memo was reasonably accurate if a trifle exaggerated, but that in criticizing the shut-off of headquarters engineers, Cook had failed to take into account that everyone at Kennedy Space Center was busy impounding data.

Judgment

Following an executive session on February 14, the commission announced its first conclusion about the accident. "In recent days," Rogers stated, "the commission has been investigating all aspects of the decision-making process leading up to the launch of the *Challenger* and has found the process may have been flawed."

The nature of the "flaw" was not immediately disclosed, but the effect of this conclusion was decisive. Rogers asked Acting NASA Administrator Graham not to include on NASA internal investigating teams persons involved in the decision process. Rogers said President Reagan had been informed of this decision.

The commission's conclusion that the launch decision-making process had been flawed was derived from testimony in executive session February 14 about the following sequence of events the night before the *Challenger* launch:[1]

---

[1] *Report of the Presidential Commission on the Space Shuttle Challenger Accident* (Washington, D.C., 1986), vol. 4, p. 599.

1. A group of Morton Thiokol engineers opposed launching *Challenger* on the 28th because of a forecast of unusually cold weather. They contended that low temperature at the launch site would reduce the resilience of the O-rings at the field joints, allow hot gas to blow by the seals and penetrate the half-inch steel casing of the booster. Accordingly, they recommended against a launch on the 28th and urged a delay until the outside temperature reached 53° Fahrenheit—an environmental condition within shuttle launch experience. The shuttle had never been launched at a lower temperature. The Air Force meteorological service was predicting a temperature of 34° F at launch time, 9:38 A.M. on the 28th.

2. This recommendation was passed to shuttle managers at the Marshall Space Flight Center during a telephone conference involving NASA and Thiokol engineers at the Kennedy and Marshall centers and Thiokol plant at Brigham City, Utah. It evoked immediate protests from Marshall executives. They challenged the data on which the no-launch recommendation was based and, in effect, called upon Thiokol engineers to prove it would not be safe to launch.

3. A recess was called during which the Thiokol engineers caucused with their management to discuss the Marshall protest. The outcome was a reversal of the no-launch recommendation, and the launch was approved by Thiokol management in writing.

4. High-level NASA officials were not informed of the Thiokol engineers' recommendation against launch nor of their concern about the effect of cold weather on the joint seals. These concerns had been treated as an internal matter by the Marshall directorate and resolved at its level.

With its request to exclude NASA personnel involved in the launch decision from the internal investigation of the accident, the commission undertook a direct, hands-on investigative role. Thus, the commissioners themselves became investigators as well as analysts of the data presented by NASA and the contractors. It was as if the members of a grand jury had gone out to look for evidence. NASA's Design and Data Analysis Task Force, although conducting its own investigation, became in fact an arm of the commission.

Commissioners were divided into four investigative panels: development and production; prelaunch activities; mission planning and operations; and accident analysis. Events now moved rapidly. On February 20, NASA headquarters announced that Rear Admiral Richard H. Truly was appointed Associate Administrator for Space Flight, replacing Jesse Moore. In that capacity, Truly took charge of the Data and Design Analysis Task Force. He promptly reorganized it in accordance with Rogers' request that it not include anyone who had been involved in the launch decision process. This action not only removed senior officials from active roles in the internal investigation but cast them in the role of witnesses.

Among those involved in the launch decision-making process were the key directors of the shuttle program: Jesse Moore, as ex–associate administrator, now transferred to the Johnson Space Center as director; Arnold Aldrich, shuttle program boss at Johnson; William R. Lucas, director of the Marshall Space Flight Center; Richard G. Smith, director of the Kennedy Space Center; Robert Sieck, launch operations chief, and James A. (Gene) Thomas, KSC launch director.

It was clear that the commission had embarked on an investigation of NASA's handling of the shuttle program as an aspect of the accident. By sidelining the agency's top launch decision-makers, the commission hoped to sidestep the contretemps of NASA investigating itself.

But the hearings in which the events of the night of January 27 were related had shown that the decision-making process had actually been worked out at Marshall, with the acquiescence of Thiokol management, at the third level of agency's management structure. Moore (Level 1) and Aldrich (Level 2) had said they were unaware of the Thiokol engineers' recommendation against the launch. Launch Director Thomas told Commissioner Hotz: "I can assure you that if we had that information, we wouldn't have launched if it hadn't been 53°."

Meanwhile, Congress entered the investigation. On February 18, Chairman Rogers and Vice Chairman Neil Armstrong appeared before the Senate Commerce Committee's subcommittee on Science, Technology, and Space. Rogers repeated his pronouncement of February 16: "Our intensive review to date has indicated that the decision-making process may have been flawed. We have not said the decision was flawed; we have said that the process may have been flawed, and we base that on testimony taken in executive session."

The semantic distinction between a flawed decision and a flawed decision-making process provided the rationale for the commission's reorganization of the investigation. It admitted that more evidence was needed to show the decision to launch *Challenger* on January 28 was a fatal blunder, but it also asserted that the commission had sufficient information to show a failure of the decision-making process; an apparent breakdown of communication between levels of command. That apparent breakdown made the agency's command structure vulnerable.

Under questioning by Senator Slade Gorton (R-Wash.), subcommittee chairman, Armstrong stated that having been operating only a short time, the commission had not yet reached any conclusion about a specific cause of the accident.

Gorton: You have concluded that there was a failure in one of those seals but not whether that was the first failure?
Armstrong: I have not concluded there was a failure in the seals.
Gorton: I gather that you are seriously investigating the effect of the extreme cold temperatures. That it is at least a suspect in having played a major role?
Rogers: One of our members [Dr. Feynman] is at Cape Kennedy trying to get precise details about the

weather. It's not as easy as you think to find out precisely what the weather conditions were. All the facts will be fully made public.

At this point, Senator Donald Riegle, Jr. (D-Mich.) suggested that senior staff members of the subcommittee who had security clearance have the opportunity to sit in some of these sessions, as listeners. He feared that executive session transcripts would not be available.

Rogers warded this off, saying: "I don't think we can do that because we've had so many requests from people wanting to sit in on the commission. After all, it is a presidential commission. Because of the separation of powers, I think we should have the option of proceeding in executive sessions."

Riegle persisted: "It seems to me that we have parallel responsibilities here, and I'm not quite sure that whatever you're learning in executive session ought not to be available to the rest of us."

Rogers remained firm. "The executive branch doesn't normally sit in on Senate hearings, private sessions. Wouldn't ask to. This is a presidential commission and we think we have the right . . . to hold executive sessions as we see fit."

Reorganization

Truly's appointment as associate administrator launched the reorganization of NASA headquarters that was to be completed when President Reagan recalled former administrator James C. Fletcher to head the agency. A native of Mississippi, Truly, 48, was a graduate in aeronautical engineering of the Georgia Institute of Technology, a Navy fighter pilot, and had spent four years in the Department of Defense's Manned Orbital Laboratory program before joining NASA in 1969 when MOL was canceled.

At NASA, Truly was a member of the support crews for all three manned Skylab space-station missions (1973–74) and a member of the support crew of the 1975

Apollo-Soyuz Test Project.[2] With Joe H. Engle, Truly comprised one of the two crews that flew the test orbiter, *Enterprise,* through approach and landing tests in 1977. Then he was the backup pilot for the first orbital flight of *Columbia;* pilot of *Columbia's* second flight; and commander of STS-8, *Challenger's* third flight.

At headquarters, he was the first astronaut to break into NASA's Level 1 executive rank. His administrative experience was principally acquired as head of the Naval Space Command that operates the Navy's satellites. He had resigned from the Corps of Astronauts in 1983 to take the post, with the rank of rear admiral. Now he was at NASA with the mission of leading NASA's investigation into the *Challenger* accident and the responsibility for reshaping the shuttle program.

As his deputy, Truly appointed James R. Thompson, deputy director for technical operations at the Princeton University Plasma Physics Laboratory. Thompson was an old NASA hand, too. Before Princeton, he had managed the development of the shuttle's main engines as associate director of engineering at the Marshall Space Flight Center.

Other key task force people were Navy Captain Robert Crippen, NASA's most experienced shuttle pilot; Colonel Nathan Lindsay, commander of the Air Force Space and Missile Center; Dr. Joseph Kerwin, former Skylab astronaut crewman and now director of Space and Life Sciences at the Johnson Space Center; and Walter Williams, special assistant to the NASA administrator, who had participated in spacecraft development since Project Mercury.

Six teams drawing engineers and technicians from the NASA centers formed in the task force. Perhaps the most visible was the Salvage Support Team headed by Colonel Edward A. O'Connor, Jr., director of operations of the 6555th Aerospace Test Group at Patrick Air Force Base, Florida. Its mission was the recovery of *Challenger* wreckage from the Atlantic Ocean.

---

[2] The Apollo-Soyuz Test Project provided the specular link-up of American and Soviet spacecraft in a demonstration of détente in orbit.

The Night of January 27

On February 25, the commission began three days of public hearings about the events of the night of January 27. At this stage, the focus of the investigation turned to the Marshall Space Flight Center and its role in the launch of *Challenger.* The hearings were held in the Dean Acheson auditorium of the State Department in Washington and televised nationally by a cable news network.

A star witness was Allan J. McDonald, manager of Morton Thiokol's solid rocket motor project. He was one of the group of company engineers who opposed the launch on the 28th because of unusually cold weather predicted for that day. The hearings probed the process in which company management overrode the objections of the engineers and recommended launch and the extent to which Marshall officials influenced that decision.

McDonald had persisted in his opposition even after his boss, Joseph C. Kilminster, company vice president, had telefaxed Morton Thiokol's recommendation for the launch to NASA.

McDonald said that he and Kilminster alternated as company chief representatives at shuttle launches and that he represented the company at KSC for the *Challenger* launch.

The day before the launch, McDonald said that he was visiting the home of Carver Kennedy, vice president of Morton Thiokol Operations in Florida, at Titusville when he received a telephone call there from Utah. The caller was Robert V. Ebeling, manager of Thiokol's solid rocket motor ignition system, who had heard a weather prediction of 18° F at the *Challenger* launch site on the morning of the 28th, McDonald said. Ebeling advised that some of the company engineers were worried about the O-rings functioning at such a low temperature. He asked for updated information on temperature projections so that the effect on the O-ring seals could be calculated.

Carver Kennedy then called the launch operations

center for temperature data. He was told, McDonald related, that a freeze was expected before midnight and that the temperature was expected to drop to 22° by 6 A.M. At the scheduled launch time, 9:38 A.M., it was expected to be 26° F. McDonald called back to the plant and relayed this information to Ebeling.

"I told him I thought this was very serious," McDonald testified. "And to make sure he had the vice president for engineering [Robert Lund] involved in this and all his people; that I wanted them to put together some calculations and a presentation of material . . . to make sure this decision [to launch] should be an engineering decision, not a program management decision."

McDonald said he then tried to reach Lawrence Mulloy, who was staying at the Holiday Inn on Merritt Island, about fifteen miles from the space center, but was unsuccessful. He then called Cecil Houston, resident manager for the Marshall Space Flight Center, at KSC and told him about the concerns of Thiokol engineers in Utah.

McDonald and Houston decided that a three-way telephone conference among engineers in east central Florida, north Alabama, and Utah should be arranged to discuss the temperature issue. Houston then agreed to set up a teleconference with a four-wire system next to his office in a trailer complex at the center, just across the road from the mammoth Vehicle Assembly Building. It was set at 8:15 P.M., EST.

McDonald said he asked the Thiokol engineers to telefax charts to Kennedy and Marshall for the display of data supporting their concerns. The data would show instances of O-ring erosion, blow-by, and charring that had been found on previous launches. It was his expectation, McDonald said, that the data would support a rationale to launch or not to launch.

McDonald then related that he arrived at Houston's trailer office at 8:15 P.M. and met Mulloy there with Stanley Reinartz, manager of Marshall's shuttle project office and Mulloy's boss. It took about half an hour for the telefaxed charts to arrive from the plant and be reproduced. When the teleconference got under way at 9 P.M., Thiokol engineers led off with a presentation of the temperature problem, based on their charts.

"They presented a history of some of the data we had accumulated both in static test and in flight tests relative to temperatures and the performance of the O-rings," McDonald said. "And [they] reviewed the history of all of our erosion studies of the O-rings in the field joints, any blow-by of the primary O-rings with soot products of combustion or decomposition that had been noted, and the performance of the secondary O-rings."

While the teleconference was going on, the seven men and women of the *Challenger* crew were resting in their quarters at the space center's administrative complex about eight miles from launch pad B. *Challenger* stood in a floodlit haze, a glowing monument in a wilderness of sand and palmetto scrub. None of the crew imagined the portents of the three-way discussion across America that would decide their fate the next morning. The persistent problems with the segment joint seals had never been discussed with the astronaut crews whose lives depended on them.

At the conclusion of the engineering presentation at Thiokol, said McDonald, the company vice president, Lund, gave the conclusions and recommendation of the group.

"The bottom line was that the engineering people would not recommend a launch below 53° F," he said.[3] "The basis for that recommendation was primarily our concern with the launch that occurred about a year earlier in January of 1985—51-C. It was our motor number SRM 15. That particular motor had a couple of field joints that not only had some erosion but they had

[3] Records at Kennedy Space Center show air temperature of 66° F on January 24, 1985 when *Discovery* 51-C was launched, but heat transfer experts at Thiokol calculated the seal temperature at 53° F when they inspected the motor at Brigham City, Utah. (Commission hearing transcript) February 25, 1986, p. 1393.)

some fairly severe blow-by of the primary seal, fairly heavy soot over a fairly large arc, very deep and very black.

"And even though we could see no measurable erosion on the secondary O-ring, it had a heat effect and by that—the sheen was gone off the O-ring seal . . . and because of that we were concerned with the launching beyond our experience base, below that temperature."

The recommendation elicited "a lot of strong comments and reactions" from several NASA officials, especially from Mulloy and George Hardy, deputy director of science and engineering at Marshall, McDonald said. He said that Mulloy, who was seated in the trailer with him, "made some comments about when will we ever fly if we have to live with that."

Mulloy was also heard to say: "My, God, Thiokol, when do you want me to launch—next April? You guys are generating new launch commit criteria."[4]

Comments were made also about plans to launch shuttles at Vandenberg Air Force Base, California where it was not unusual to have morning temperatures below 53° F, McDonald said. He added that he had heard Hardy say from Huntsville that he was "appalled" at the Thiokol engineers' recommendation.

"But he [Hardy] also said he would not fly without Thiokol's concurrence," McDonald said.

McDonald said he heard Reinartz remark that he was under the impression that the motors were qualified from 40° to 90° F, and that the 53° recommendation was inconsistent with those parameters. Cecil Houston observed that morning temperatures at the space center probably wouldn't reach the fifties until Thursday, January 30.

The critics at Marshall challenged Thiokol engineering conclusions on temperature effects on O-ring erosion and blow-by. There were data showing that in the late fall of 1985 (October 30), the seals in motor 22

---

[4]Commission hearing transcript February 26, p. 1540.

---

showed blow-by but no erosion following the launch of *Challenger* 61-A, McDonald said. The air temperature was 85° F, according to KSC records.

"So there was some concern," McDonald said, "that the data was inconclusive and also that we had some motors that were static-tested as low as 36° and it [development motor no. 4] showed not only no O-ring erosion but no blow-by." However, he added, another senior engineer, Roger Boisjoly at Thiokol, asserted that there had been a significant difference in the volume of soot that had passed the primary O-ring on motor 15—the cold launch—and motor 22, the warm launch. The cold launch effect was a much larger arc of soot between the two O-rings, and the soot was much blacker than on motor 22.

Moreover, McDonald continued, the motor 15 blow-by penetrated as far as the secondary O-ring, which showed heat effect. It was Boisjoly's contention, McDonald said, that the difference in the blow-by of the seals in these two motors was due to temperature—the lower the temperature, the greater the blow-by and the darker and denser the soot.

Moreover, McDonald continued, the static-test history that Marshall executives had cited was invalid in terms of temperature effect because the static-test motors were kept indoors at 70–72° F until exposed to ambient weather outside about six hours before the test. At that time, he said, he had commented that the lower temperatures were "in the direction of badness for both O-rings" because it slowed down their timing function. The effect was much worse for the primary O-ring because the leak check forced the primary O-ring on the wrong side of the O-ring groove, while at the same time it forced the secondary O-ring in the proper direction. These effects should be weighed, he said, in making an evaluation of the recommended temperature.

McDonald displayed a chart showing how the O-rings were supposed to work. Seating of the primary O-ring was critical during the first 0.6 seconds of the ignition transient, he explained. That process was affected by cold weather. Grease in the sealing area would become viscous and stiff, and the O-ring itself would become stiff.

McDonald's analysis confirmed the results of Feynman's icewater experiment. "We knew that cold temperatures shrank the O-ring some," the senior Thiokol engineer continued, "and from our resiliency tests which are tests that basically showed how the O-ring responds when you have it under some pressure and release that load . . . it shows that it gets cold and stiff. It doesn't want to respond very well."

Cold also affected the secondary O-ring, McDonald said. He characterized it as a redundant seal until the metal parts between the tang and clevis began to rotate as the steel case inflated under engine gas pressure. Once rotation started, and especially if there was reduced resiliency, the secondary O-ring might not be redundant.

With this explanation, McDonald continued his account of the teleconference the night of January 27. Shortly after Thiokol engineers completed their presentation, he said, Thiokol was queried by Marshall about the no-go recommendation. "I believe Joe Kilminster was asked himself what his recommendation would be, since it was engineering that recommended not flying, and he said he would not go against that recommendation," McDonald said.

But, he added, because of the controversy over the effects of temperature, "We were asked to reevaluate the data, and the people in Utah said they would like to have a caucus for five minutes and go off the line. Lund had presented the recommendation, and no one recommended launch."

During the recess, McDonald waited in the KSC quarters with Reinartz, Mulloy, Houston, and Jack Buchanan, an engineer. McDonald said he suggested a late afternoon launch when it would be warmer. He said he had been told that an afternoon launch had been considered but rejected because of weather or visibility problems at one of the transatlantic abort landing sites.

About a half-hour later, he continued, the Thiokol

people came back on the line. "I believe it was Joe Kilminster who came on the line and said that even though we had some concerns about the lower temperatures that we would recommend that they proceed with launch based on the fact that we felt the temperature data that we had was not totally conclusive. He outlined some rationale why we felt it was safe to proceed."

Hardy, the senior Marshall official on the line, then asked Kilminster to put the recommendation in writing, sign it, and make sure to get it to the Kennedy Space Center by early morning. McDonald said that he would not sign the recommendation, that it would have to come from the plant; Kilminster agreed to draft a statement and send it down.

The message from Kilminster, titled "MTI [Morton Thiokol, Inc.] Assessment of Temperature Concern on SRM-25 [51-L] Launch," was sent by telefax to the Marshall and Kennedy space centers at 9:45 P.M. Mountain Standard Time (11:45 EST), January 27. It said:

Calculations show that SRM-25 O-rings will be 20° colder than SRM-15 O-rings [on the *Discovery* launch of January 24, 1985]. Temperature data not conclusive on predicting primary O-ring blow-by. Engineering assessment is that: Colder O-rings will have increased effective durometer ("harder"); "harder" O-rings will take longer to "seat." More gas may pass primary O-ring before the primary seal seats (relative to to SRM-15). Demonstrated sealing threshold is three times greater than 0.038" erosion experienced on SRM-15. If the primary seal does not seat, the secondary seal will seat. Pressure will get to secondary seal before the metal parts rotate. O-ring pressure leak check places secondary seal in outboard position which minimizes sealing time. MTI recommends STS 51-1 launch proceed on 28 January 1986. SRM-25 will not be significantly different from SRM-15.

The message was signed by Joe C. Kilminster, Vice President, Space Booster Programs, Morton Thiokol, Inc.

McDonald recalled saying that if anything happened to this launch, "I sure wouldn't want to be the

O Calculations show that SRM-25 O-rings will be 20° colder than SRM-15 O-rings

O Temperature data not conclusive on predicting primary O-ring blow-by

O Engineering assessment is that:

    O Colder O-rings will have increased effective durometer ("harder")

    O "Harder" O-rings will take longer to "seat"

        O More gas may pass primary O-ring before the primary seal seats (relative to SRM-15)

            O Demonstrated sealing threshold is 3 times greater than 0.038" erosion experienced on SRM-15

    O If the primary seal does not seat, the secondary seal will seat

        O Pressure will get to secondary seal before the metal parts rotate

            O O-ring pressure leak check places secondary seal in outboard position which minimizes sealing time

O MTI recommends STS-51L launch proceed on 28 January 1986

    O SRM-25 will not be significantly different from SRM-15

Joe C. Kilminster (signature)

JOE C. KILMINSTER, VICE PRESIDENT
SPACE BOOSTER PROGRAMS

MORTON THIOKOL INC.
Wasatch Division

INFORMATION ON THIS PAGE WAS PREPARED TO SUPPORT AN ORAL PRESENTATION
AND CANNOT BE CONSIDERED COMPLETE WITHOUT THE ORAL DISCUSSION

Copy of the message sent by telefax to NASA by Morton Thiokol, Inc. Vice President Joe C. Kilminster recommending that *Challenger* be launched on January 28, 1986. (*Commission Report*, vol. 4, p. 753)

person that had to stand in front of a board of inquiry to explain why we launched this outside of the qualification of the solid rocket motor on any shuttle system." He told the commission that he asked that the recommendation to launch be reconsidered for three reasons: first, concern about cold O-rings; second, the fact that booster recovery ships were in a "survival mode" battling 30-foot seas and 50-knot winds that night and would not be in any position to support an early morning launch; third, the formation of ice on the launch pad.

"I was told, you know, these really weren't my problems and I really shouldn't concern myself with them," McDonald said. "But I said all of these together should be more than sufficient to cancel the launch if the one we discussed earlier wasn't."

During testimony from McDonald and subsequent witnesses about the temperature effect on the O-rings, it became apparent to observers that this issue had never been resolved. Testimony conveyed an impression that it was an experimental rocket and its performance that were being discussed, not one that had been operating for years and had been touted as the product of a mature technology.

Later that evening, Mulloy conferred by telephone with Arnold Aldrich, the shuttle manager at Houston, who was concerned about recovering the boosters at sea in view of the rough sea states and high winds. McDonald said that he heard Mulloy say that he expected recovery of the boosters despite high seas, inasmuch as they were equipped with locator beacons, but there was a high probability that the parachutes and frustrums (nose structures) would be lost. Aldrich asked Mulloy about the value of the items that might not be recovered and "Larry gave him some number close to $1 million," McDonald said.

Aldrich, at Level 2 in the NASA command chain, had launch decision-making authority. He was concerned about ice damage to the orbiter heat shield during the launch, according to McDonald.

"I didn't hear anything discussed about the O-ring seal problem," McDonald added.

"But you're not sure of that?" asked Rogers.

McDonald replied that he was not, because he had left the room for a while to pick up the telefax from Kilminster.

Aldrich testified later that he had not been informed of the no-go launch recommendation by Thiokol engineers and its reversal. Mulloy later said that he had not mentioned it in the evening's conversation with Aldrich.

Rogers asked: "How could the Thiokol people change their minds about the launch based on inconclusive data?" McDonald said he couldn't explain it. Rogers then asked McDonald if he or the company were under pressure to reverse the engineers' recommendation.

"There was no doubt in my mind—I felt some pressure," McDonald replied.

Not only the launch recommendation but the roles of customer and contractor were reversed in this situation, he said. In the flight readiness reviews that always preceded a launch, the contractor was required to explain why it was safe to fly to a critical customer

(Marshall). On 51-L, the contractor was challenged to explain why it was unsafe to fly, and now the challenge came from the customer.

"I was surprised at this particular meeting. . . . the contractor always had to stand up and prove that his hardware was ready to fly," McDonald said. "In this case, we had to prove it wasn't." This reversal also surprised observers. Commissioner Arthur B. C. Walker, Jr. of Stanford University asked the witness: "Did anyone from NASA explicitly ask for reconsideration of the decision not to launch, or did reconsideration occur because of the negative remarks and comments on the decision?"

"Well, I think it was the latter," said McDonald. "I can't fully recall whether they directed us to do that or not but they had concluded that the temperature data was inconclusive and I don't know whether we volunteered to reassess it or whether they said we needed to."

Commissioner David C. Acheson, Washington attorney, asked whether McDonald had considered bringing his concerns about the final recommendation to the personal attention of Jesse Moore, Arnold Aldrich, or William Lucas. McDonald said that he was sure that the issues were brought to their attention "because that is the way things go." He said that he assumed that Mulloy and Reinartz had passed along his concerns.

"Was the final decision to launch from Thiokol engineering or management?" asked Commissioner Robert Hotz, retired aerospace journal editor.

"I guess I would have to characterize it as a management decision," said McDonald.

Astronaut Sally Ride, the only woman member of the commission, questioned McDonald about the criticality 1 designation of the field joint primary O-rings. McDonald said he first learned about the criticality 1 classification in August 1985 during an O-ring presentation by Thiokol and Marshall to NASA headquarters. "I was unaware of it myself that we had such a condition, that we were flying crit 1 on that part of the hardware," he said.

He went on to say that there had always been adequate squeeze on the O-rings from the hardware even when rotation occurred and that it had maintained redundancy in the secondary seal, with one exception, the fourth flight of *Columbia.*

He explained: ". . . there was only one time where we had a tolerance stack-up of a joint that would have fallen from criticality 1-R to 1, and that was on STS-4. We never had one since."[5]

Columbia had been launched on its fourth and final test flight June 27, 1982, with Thomas K. Mattingly II, commander, and Henry W. Hartsfield, Jr., pilot. Booster performance was poorer than expected, and the ascent trajectory was lower and slower than planned. A problem in the seals was suspected but could not be ascertained because the booster parachutes failed and both boosters sank to the bottom of the sea before retrieval ships could reach them.

Columbia nevertheless reached its 184-mile orbital altitude with four firings of its space engines. The balance of the mission was uneventful, and when *Columbia* landed at Edwards Air Force Base on the morning of July 4, President Reagan and Nancy Reagan were on hand with 200,000 spectators to greet the crew and hail the completion of the shuttle test-flight program. NASA proudly announced that from then on, the shuttle was an "operational" transportation system.

The loss of the two boosters was explained as a malfunction of the parachute descent system, and the incident was virtually forgotten until McDonald recalled the STS-4 tolerance stack-up. Its effect on the seals would ever be speculative, however, because the boosters were never recovered.

Rogers asked if it was not a fact that since December 1982, the shuttle had been launched with field joint seals classified as criticality 1, meaning that if the primary seal failed, a catastrophe could have resulted because there was no redundancy. McDonald replied

[5] A tolerance stack-up of a joint refers to a probable misalignment during stacking of the motor case segments that could have prevented the secondary O-ring from sealing.

that he guessed that was how criticality 1 would be interpreted, but, recognizing the actual hardware, he said he did not believe that was true.

His and later testimony expressed faith that if the primary O-ring somehow failed during the ignition transient, gas pressure would seat the secondary O-ring and keep it seated against dislocation by joint rotation. On that process, most Thiokol and Marshall engineers tended to agree, although there was dissent on each side.

"But just to be clear," said Dr. Ride, "what the critical items list says is that the primary O-ring is a criticality 1 and you're not allowed to consider the secondary O-ring as a backup to that."

"That's true, Sally," said McDonald. "That's absolutely correct. That is what is meant."

Rogers: And that was known by everyone, I assume, who was working on the program or most of the top people working on the program. Would that be true?

McDonald: Well, I kind of thought I was one of the top people working on the program and I didn't know that until August 1985 when I put that presentation [on O-ring problems] together.

Feynman: Suppose the pressure is increasing for some reason in the primary seal. The primary seal begins to erode, and by the time the pressure gets to some figure like 600 or 700 pounds per square inch, which I think is just below maximum operating pressure, it finally erodes all the way through so that the gas can pass through the primary seal. Would you think there was a reasonable probability that the whole thing would fail because the rotation [of the joint] by that time was enough so that the secondary seal can't hold it?

McDonald: That was our assessment in August—that there was such a reasonable probability.

Feynman: That you first knew or thought of in August 1985?

McDonald: That is correct.

He added that in the fall of 1985, some of Thiokol's top engineers attended a conference of the Society of

Automotive Engineers "to get help from the whole seal industry and the SAE about the field joints."

A Substantial Margin

The next witness, Jerald E. Mason, senior vice president of the Wasatch Division operations at the Morton Thiokol plant in Brigham City, said his concern had been aroused the day before the launch by information that a 34°F temperature was predicted at launch time, 9:38 A.M.

"The concern was whether the cold and the stiffness of the O-ring and the grease would delay the movement of the primary O-ring so that it would not seal properly and there would be blow-by and perhaps damage to the secondary O-ring," he said. "We looked at the erosion history . . . to see what correlation there was with temperature, and there really wasn't any correlation. . . . we had blow-by in both 75 and 53°, but it was much more severe at 53°, and so it was thought that that was evidence that the cold did in fact affect the performance of the O-ring."

Mason said that data on the hardness of the O-ring was studied as an indication of how stiff it was and how difficult it might be for the O-ring to move. It was determined that a 20-degree drop in temperature, from 50 to 30°, would increase the hardness by about 10 percent, he said. Early in the program, a 90-durometer (measure of hardness) O-ring had been used in some hydrostatic tests, but "we ended up with the 80-durometer O-ring." He added that it was pointed out (during the discussion) "that we had some experience with that O-ring that was harder."

Feynman: What you mean by hardness is, it's difficult to squash it into the tiny crack that it is supposed to go into? What is the feature of hardness? It is not resiliency—it is the difficulty of pushing it into a corner.

Mason: Well, it was thought that stiffness might make it move more slowly, moving across the gap, and then it would be harder to extrude it in. However, once it sealed, being slower to extrude per se wouldn't be a problem once it sealed. The points we're talking about were the points that were discussed back and forth that night.

Mason said that in looking at the history of previous tests and flights, "we discussed the fact that we had run a subscale [small motor] blow-by test at 30°, which had shown no blow-by." He cited a test where a cut of 125 thousandths of an inch would still let the O-ring seal and another test that showed that three times as much O-ring erosion could be tolerated as had been seen on the worst examples of that effect on 51-C.

"We felt we had a substantial margin," he said.

Dr. Feynman asked if Mason had a calculation that showed "you couldn't get more erosion than the amount you got on 51-C." Mason said he didn't know.

Feynman: What made you think that 51-C was the maximum erosion that you could possibly expect?

Mason: We felt that the factor of three times was not likely to be exceeded.

Ride: I guess what I'm really trying to understand is whether you really had the engineering data or an engineering analysis to back up the decision that this criticality 1 was safe to fly at those temperatures.

Mason: The reason for the discussion was the fact that we didn't have enough data to quantify the effect of the cold, and that was the heart of our discussions.

Commissioner Joseph F. Sutter, vice president of the Boeing Commercial Airline Company, commented: "Your engineers with the data they had and with the concern with the temperature did reach a tentative conclusion at least that, why not wait for at least a temperature that had already been flown. . . . I am extremely puzzled why a NASA person could disagree and ask you to review the decision your engineers had reached, since theirs was the responsibility for the design."

Mason explained: "It was difficult to say that 53° was exactly the temperature that you ought to fly at. . . . In any event we expected the primary to seat . . . because we had some tolerance of its ability to erode and still seat and seal. That was our thought process."

From No-Go to Go

The process by which the no-go recommendation of the Thiokol engineers was reversed by management was described by Roger Boisjoly, a senior company engineer on the task force the company had set up in the summer of 1985 to improve the seals.[6] He appeared before the commission with Arnold Thompson, a company structures supervisor.

Boisjoly observed the disassembly of solid rocket motor 15 (from 51-C). This was the first time the primary O-ring in a field joint had been penetrated by hot gas, he said. By contrast, the blow-by in motor 22 on 61-A, the *Challenger* flight launched October 30, 1985 at an ambient temperature of 84°, was much lighter in color. "This told me that temperature was indeed a discriminator." Lower temperature was away "from the direction of goodness."

Rogers asked if the no-go recommendation was unanimous.

"Yes," said Boisjoly. "There never was one positive pro-launch statement made by anybody."

Thompson, Boisjoly's colleague, agreed. "Particularly in the caucus," he said, "Roger and I were the only people who expressed our views."

Boisjoly said that the caucus was started by Jerry

Mason stating that a management decision was necessary. Boisjoly and Thompson continued to speak against the launch. "Again, I brought up the point that SRM-15 had a 110-degree arc of black grease while SRM-22 had a relatively different amount which was less and wasn't quite as black."

Commissioner Walker asked: "At this point did anyone else speak up in favor of the launch?"

"No, sir," said Boisjoly. "Nobody said a word. After Arnie [Thompson] and I had our last say, Mr. Mason said we have to make a management decision. He turned to Bob Lund and asked him to take off his engineering hat and put on his management hat.

"From that point, management formulated the points to base their decision on. There was never one comment in favor, as I have said, of launch from any engineer or other nonmanagement person in the room, before or after the caucus. I was not even asked to participate in giving any input to the final decision charts."

Kilminster presented the final chart that was the rationale for launching, Boisjoly said. It was handwritten on a note pad and he read from the pad, Boisjoly said, adding: "It was clearly a management decision from that point. I left the room feeling badly defeated but I felt I really did all I could to stop the launch. I felt personally that management was under a lot of pressure to launch and that they made a very tough decision, but I didn't agree with it."

Next, Robert Lund, vice president of engineering, Joe C. Kilminster, vice president of the shuttle project, and Brian Russell, engineer, were called as Morton Thiokol witnesses. Lund at first had sided with the engineers but had accepted management's pro-launch position. Rogers asked him: "How do you explain the fact that you changed your mind when you changed your hat?"

Lund explained: "We have dealt with Marshall for a long time and have always been in the position of defending our position to make sure we were ready to fly, and I guess I didn't realize until after that meeting,

[6] The task force was organized in response to a memorandum from Boisjoly warning of disaster if something wasn't done about the O-ring seals.

and after several days, that we had absolutely changed our position from what we had been before.

"But that evening, I guess, I had never had those kinds of things come from the people at Marshall. We had to prove to them that we weren't ready. And so we got ourselves in the thought process that we were trying to find some way to prove to them it wouldn't work and were unable to do that. We couldn't prove absolutely that the motor wouldn't work. . . .

"As a result of that telecon, I gave the charts and made the recommendation that we wait until the motor got to 53° and, of course, you heard the story of what happened after that. And so I think that as Marshall pointed out—I think Mr. Mulloy pointed out—he said, you know, the data is just not conclusive at all—and it wasn't, because we had a low-temperature motor and a high-temperature motor and we had ten motors in between that showed nothing.''

Commissioner Sutter observed that it appeared that erosion and blow-by had been occurring more frequently "later than sooner." Lund admitted: "We haven't been able to identify those parameters that are causing that more pronounced effect."

"Why didn't you just tell them [NASA] it's our decision and this is it, and not respond to the pressure?"

"As a quarterback on Monday morning," Lund replied, "That is probably what I should have done."

Commissioner Armstrong asked if there were factors other than temperature that might be involved in the seal problems.

"That is correct," said Lund. "So we don't know what the effect of temperature was. There was no full-scale motor fired below 40°. The development motors were fired from 40 to 84° and the qualification motors from 45 to 83°.''

Commissioner Walker: So then 40° was the temperature limit for the O-rings? [The question referred to the fact that *Challenger* was launched at ambient temperature of 36° F.]

Lund: In the full-scale qualification program, that is correct.

Walker: So when it was predicted that the O-ring temperature at the launch of 51-L was going to be 29°, the O-ring was outside of the qualification temperatures by some degrees?

Lund: That is correct.

Let's Take a Chance?

Walker: Then how could you make a recommendation to launch if you were 10 degrees outside of your qualification?

Lund: Our original recommendation, of course, was not to launch.

Walker: . . . but your final recommendation was to launch.

Rogers told Lund that he must have known that NASA would not have launched if he had voted against it. Lund replied that he couldn't predict what NASA would do.

Rogers: But you knew that was the reason they asked you to reconsider—that is why you had the five-minute recess. . . .

Lund: That's a fair statement, yes.

Rogers: Now, knowing that and knowing that the safety of the crew was involved, and knowing your own people, the engineers that you respected, were still against the launch, what was it that occurred in your mind that satisfied you to say, okay, let's take a chance?

Lund: Well, I didn't say take a chance because I felt that there were some rationale that allowed us to go ahead.

Rogers: Maybe that isn't fair. But what was it that occurred in your mind that caused you to be willing to change your mind?

Lund: I guess one of the big things that we really didn't know whether temperature was the driver or not. We couldn't tell. We had hot motors that blew by and cold motors that blew by and some that did not. The data was inconclusive. And so I had trouble justifying it in my own mind and saying, by golly, temperature is a factor.

Rogers then turned to Kilminster, who had signed the company statement recommending launch. "How can you say you changed your mind when you say temperature and data [are] inconclusive on predicting O-ring blow-by? Did you have a feeling that the burden of proof was on you to show that it wasn't safe?"

Kilminster denied that NASA had put pressure on him to change his mind and repeated that the test data had seemed inconclusive.

No Temperature Criteria

The commission continued to pursue the temperature question the next day. Mulloy and Hardy of the Marshall Space Flight Center were the first witnesses. Mulloy took the position that the low-temperature limit of launch commit criteria was 31°, not 40°, as previously mentioned, a parameter that had been established early in the shuttle program. The 40-degree limit referred to the bulk temperature of the propellant, which was not as susceptible to changes in outside weather temperature as were the joint seals. Mulloy contended that the seals were actually redundant during the ignition transient and, despite the reclassification of the field joint primary seal as a criticality 1 item, redundancy would be lost only under worst-case conditions.

Rogers asked if the weather on the 28th and the recommendation of the Thiokol engineers did not represent a worst-case condition. Mulloy replied it did not seem that way to him.

Commissioner Armstrong asked Mulloy: "would you expect that after full flight pressurization of the motor you would have a secondary seal? Do you think on these joints you had a crit 1 or 1-R?"

Given the temperatures and known resilience of the O-rings, said Mulloy, "it would be my judgment that with the resiliency data that was presented on the 27th, that would be a condition where the secondary seal may not function."

"After joint rotation," added Hardy.

"After joint rotation," confirmed Mulloy.

"So just to repeat that," said Armstrong, "in this case we might have a single seal failure, namely the primary seal failure, after motor pressurization that could cause a problem of the kind we are investigating?"

"Yes, sir," agreed Mulloy. "That is the condition recognized in the critical items list."

Mulloy then explained his thinking on this question. Harking back to the teleconference on the 27th, he said: "I asked Mr. Kilminster. . . . for his recommendation for 51-L. He stated that based on the engineering recommendation, he could not recommend launch. But where I was coming from is, we had been flying since STS-2 with a known condition in the joints that was duly considered and accepted by Thiokol; it was accepted by me and it was accepted by all levels of NASA management through the flight readiness review process, through special presentations that we had put together and provided up here to the headquarters people."

Commissioners Walker and Armstrong both said they understood that temperature criteria Mulloy had mentioned covered the entire vehicle. Why was there no specific launch commit temperature criterion for the O-rings?

"Why hadn't something bubbled up through the system that would indicate a more well-defined constraint on launch?" Armstrong asked.

" . . . There just wasn't any great concern expressed about temperature," said Mulloy. "There was no focus on launch commit criteria for the joint."

Commissioner Hotz said the testimony indicated to him that the risk of low temperature on the O-rings was not transmitted to the highest levels of NASA. (Mulloy, Reinartz, and Hardy at Marshall represented Level 3 in the operations chain of command.) Hotz wanted to know if the discussion about temperature effect on the seals had been transmitted to Level 2, represented by

Aldrich at Houston, or to Level 1, represented by Moore at headquarters.

Mulloy explained: "I did not discuss with Mr. Aldrich the conversations that we had just completed with Morton Thiokol." He added that the information was not transmitted beyond the director of Marshall, William Lucas.

"Could you explain why?" asked Rogers.

At the time, Mulloy said, he considered the matter a Level 3 issue, since there was no violation of launch commit criteria. "There was no waiver required in my judgment at that time and still today. We work many problems at the orbiter, the SRBs, and the external tank level that never get communicated to Mr. Aldrich or Mr. Moore. It was clearly a Level 3 issue that had been resolved."

Chairman Rogers took another tack. "Do you remember any other occasion when the contractor recommended against launch and that you persuaded them they were wrong and had them change their mind?"

"No, sir," said Mulloy.

Rogers suggested that inasmuch as their contract was coming up for renewal, the Morton Thiokol people were under "a lot of commercial pressure to give you the answer you wanted and they construed from what you and Mr. Hardy said that you wanted them to change their minds."

"I cannot conceive how Thiokol felt any pressure for the renewal of their contract," Mulloy replied. "Because they are our sole source for solid rocket motors at this time, and that contract was going to be renewed. There was no alternative, given the mission model, so that certainly wasn't a pressure factor."

Rogers: " . . . they were concerned NASA might be looking for another contractor."

Major General Donald Kutyna, commission member, intervened. "Larry, I think that we're talking about the dual source, and you've got responses from contractors due the 14th of March. And there is some leeway as to how much you're going to buy from Thiokol versus the dual source. A minimum of six from the dual source?"[7]

"In the solicitation of interest that is on the street, yes," said Mulloy. He added that Morton Thiokol had no incentive to take risks. They were subject to a $10 million penalty if an SRM caused a criticality 1 failure, he explained, and the penalty would escalate with the loss of mission success fee. "I cannot conceive of how they would allow or think that NASA could pressure them into making an unsafe decision."

"If this were an airplane," said General Kutyna, "an airliner and I just had a two-hour argument with Boeing on whether the wing was going to fall off, I think I would tell the pilot. At least mention it. Why didn't we escalate a decision of that importance?"

Mulloy said they did. Reinartz, his manager, was at the meeting, and early the next morning they told Director Lucas about it.

"But this is not in the launch decision chain," said Kutyna.

"No, sir," agreed Mulloy. "Mr. Reinartz is in the launch decision chain, though."

Kutyna: And he is the highest level in that chain?
Mulloy: No. Normally we would go from me to Mr. Reinartz, to Mr. Aldrich, to Mr. Moore.

Reinartz reiterated that Marshall Center Director Lucas had been informed about the initial Thiokol concerns and the final Thiokol launch recommendation.

"If I could use an analogy," said Kutyna, "if you want to report a fire, you don't go to the mayor. Why didn't you go up the chain?"

Reinartz: I did not perceive any clear requirement for interaction with Level 2, as the concern was worked and dispositioned with full agreement among all responsible parties to that agreement.
Hotz: Mr. Reinartz, are you telling us that you in fact

---

[7] This referred to speculation that NASA might buy six boosters from another contractor.

are the person who made the decision not to es-calate this to a Level 2 item?

Reinartz: That is correct, sir.

Rogers cleared his throat and looked around the room. "In the Navy," he said, "we used to have an expression about going by the book and I gather you were going by the book. But doesn't that process require some judgment? Don't you have to use common sense? Wouldn't common sense require that you tell the de-cision-makers about this serious problem that was dif-ferent from anything in the past?"

Reinartz replied that he believed that "the Thiokol and Marshall people had fully examined that concern and that it had been satisfactorily dispositioned." Hardy said that he had concluded that "temperature at the levels that we were talking about was not dominant in the functioning of that joint, and therefore there was no increased flight risk."

But in spite of that conclusion, he said, "I was fully prepared and so stated that I would accept a recom-mendation of Thiokol, or the opposite of that, I would not go against the recommendation of Thiokol."

Rogers asked: "Now suppose that Mr. Kilminster

had said to you, I am sending the telefax and this represents management's decision and there are three or four of us, but all the engineers are still opposed to the launch. How would you have reacted to that?"

Hardy stated bluntly: "I would not have accepted it."

Citing this testimony, Rogers told Reinartz: "So there again it seems to me at least that there was a failure of the process. And you relied on the telefax from Mr. Kilminster; and Mr. Hardy and others didn't realize that all the engineers at Thiokol were against the launch, even then. So that information never got to you and it never got to Mr. Moore or Mr. Aldrich."

"You are correct, Mr. Chairman," said Reinartz.

In its final report, the commission stated that it was troubled by "what appears to be a propensity of man-agement at Marshall to contain potentially serious prob-lems and to attempt to resolve them internally rather than communicate them forward." The commission concluded that Thiokol management recommended the launch "contrary to the views of its engineers in order to accommodate a major customer."

6. Ice

Was there political pressure on NASA officials to launch *Challenger* as early as possible after a sequence of frustrating delays? Did that account for the decision to fly on the coldest launch day in NASA's experience?

It was rumored that the "administration" was anxious to have *Challenger* in orbit on February 4 when President Reagan was due to deliver his State of the Union message. This rumor circulated not only in Washington but abroad.[1] It suggested that plans had been made for a live communications hookup with *Challenger* during the broadcast of the message. The Rogers commission investigated the rumor but reported it found no evidence that any such plans had even been considered.

However, the fact that pressure existed to get *Challenger* off the pad is indisputable. The obvious source was the 1986 launch schedule, which clearly had overextended the space agency's resources. The immediate focus was the planet Jupiter, target of two scientific missions that, as mentioned earlier, had to be launched on a timetable dictated by a narrow launch window.

Another source was the news media. They tended to react as a goad to the agency when launches were delayed, especially for minor malfunctions or for weather uncertainty. Months after the accident, Senator John H. Glenn (D-Ohio) recalled hearing a television network correspondent jibe at NASA during a launch delay with the query: "When is that turkey going to become an eagle?" Don't think that kind of thing doesn't get to people, Glenn commented.[2]

When he testified before the Rogers commission on February 27, Jesse Moore, then associate administrator for space flight, cited pressure problems. At the L (for launch) minus one day review, he said, "we were all set to go," and then concern about the weather resulting from an unfavorable forecast persuaded mission management not to launch on the 26th. This decision was taken, he said, despite the arrival of dignitaries from the Peoples Republic of China, members of Congress, and a planned stopover to the see the launch by Vice President Bush.

"Nevertheless, we decided to scrub the launch as a result of the weather forecast," Moore said. "To my knowledge, no one had any political pressure whatever to get the launch off."

But, he added, the space agency had been "roundly criticized in the press as a result of the flight just prior to this about the multitude of delays starting in the December 20 launch attempt. We shut down the week of the holidays to give our team a rest and so forth, and then we had four or five additional scrubs before we finally got it launched, and we also waved· off [the landing] three times at Kennedy Space Center."[3]

Following the *Challenger* launch postponement of

---

[1] The author was queried by telephone by a radio commentator in New Zealand about rumors of pressure from the White House.

[2] Comment on "This Week" with David Brinkley, June 8, 1986, by Senator Glenn.

[3] Moore referred to *Columbia* flight 61-C, initially scheduled for launch December 18, 1985 and postponed serially to December 19, January 6, 7, 9, 10, and 12, 1986 because of mechanical problems and bad

January 26, the mission management team met with Lieutenant Colonel Edward F. Kolczinski, commander of Detachment 11, 2d Weather Squadron at Patrick Air Force Base, that afternoon. He reported that his forecast showed a cold front with very high winds on the way. It would be passing through the Cape Canaveral–Merritt Island area Monday morning, January 27 and would create the potential for strong crosswinds that would endanger an emergency abort landing at the space center's runway. Temperatures were expected to fall to the mid to upper twenties, he testified.

Still, the launch was made ready for the 27th, and the crew boarded in the morning. Then there was a delay for the repair of a faulty latch on the crew hatch. While it was being fixed, the predicted high winds arrived. They exceeded the 15-knot limit for a safe return to launch site abort. The launch was scrubbed.

The forecast for the 28th was clear and cold, with decreasing winds. Overnight, the temperature fell to 24°F, a severe freeze for the space coast (at Titusville). Unaware of the temperature concerns of the Thiokol engineers, launch directors did not regard the freeze per se as a constraint. Until the shuttle era, a hard freeze was a rarity. During the Mercury, Gemini, and Apollo era, minimum winter temperatures in the Titusville–Cape Canaveral area averaged 48° to 50°F.[4] January freezes hit the area in a cooling sequence in 1977, and then in 1981, 1982, 1983, 1985, and 1986. The sequence strangely parallelled five winter freezes in the 1890s.[5]

Data compiled by the Marine Resources Council of East Central Florida show that more freezes have occurred in this region in the late 1970s and the 1980s than at any other time in the twentieth century. The parallel with the 1890s suggested a cycle to council researchers. The research focused on the coastal region between the Atlantic Ocean and the Indian River, a 120-mile lagoon extending from the Titusville-Cape area southward to Stuart in Martin County. By comparing 30-year averages from 1931–60, 1941–70, and 1951–80, the study concluded that mean temperatures in the area dropped 1 to 2 degrees in 20 years. While summer temperatures showed less than 1 degree cooling, winter mean temperatures dropped 1.5 degrees at Titusville and 2.1 degrees at Fort Pierce, 60 miles south. Council researchers characterized the 20-year cooling trend and the severe freezes associated with it as "dramatic."

This climate variation of the 1980s imposed an environmental hazard on launching the shuttle that remained unrecognized until 1985 when the effect of low temperature on the booster joint seals became apparent. Even then, air temperature at the launch site was not considered as a safety parameter by mission management. It was outside the totality of launch experience at the Cape and at the Kennedy Space Center launch sites. Lightning in thunderstorms, rain, high winds near the ground and aloft—all were constraints, but not ambient temperature. Yet "killer" freezes had damaged the citrus crop severely in the 1980s.[6]

Colonel Kolczinski related that he had given an in-person briefing to the mission management team during the afternoon of the 26th. "At that time," he said, "we talked about the cold front system passing through, strong winds for the Monday morning [27th] launch time period, potential crosswinds. As an outlook which we give . . . we always give an outlook for the day after,—we indicated that once the cold high had set in in the Florida area, that we did anticipate having some

---

weather. The mission carried U.S. Representative Bill Nelson (D-Fla.), chairman of the House Space Science and Applications subcommittee, as a payload specialist. He praised NASA for its caution.

[4] Chart prepared by Fred Doehring, Marine Resources Council, Florida Institute of Technology, Melbourne.

[5] Fred Doehring, Diane D. Barile, and Karen A. Glatzel, Marine Resources Council. *Climate and Climate Trends in the East Central Coastal Region of Florida.* Paper presented at the Conference on Climate and Water Management, Asheville, N.C., August 4–7, 1986, published by the American Meteorological Society, Boston.

[6] Killer freezes in December 1983 and January 1985 cost Florida 185,000 acres of citrus, and a total of 400,000 acres were lost in six years, according to the Florida Crop and Livestock Reporting Service (*Orlando Sentinel,* September 12, 1986, p. 1).

colder temperatures. At that time, we were forecasting the mid twenties to the upper twenties."

Following the scrub on the 27th, the meteorologist said, "we presented again a forecast of clear conditions, basically, good winds, no precipital kind of weather, but we did indicate that we would have a little colder temperature than we had predicted before."

The ice inspection team visited the launch pad three times on the 28th. Members who appeared before the commission were Billy K. Davis, senior test representative for the external tank, and Charles Stevenson, a veteran pad inspector. They said their task was to look for debris that might be blown about by engine exhaust at launch. The orbiter was vulnerable to debris impacts that could crack the brittle ceramic tiles forming its heat shield and expose the vehicle and crew to high heat during reentry into the atmosphere.

The most obvious debris threat was the ice on the launch pad. A year earlier, the launch of *Discovery* 51-C had been scrubbed on January 23, 1985 because of it. *Discovery* was launched the next day at 53°F (booster seal temperature calculated from exterior case temperature). As related earlier, resulting booster seal damage at that temperature led Thiokol engineers to recommend against launching *Challenger* at any lower temperature.

The inspection group was designated by Stevenson as the ice/frost team because of its preoccupation with pad ice. The focus of its three inspections during the countdown was to assess the threat of shuttle damage by ice on the fixed service structure, the rotational service structure, and the mobile launch platform on which the shuttle stood.

A freeze protection plan, used the previous January on pad A, was carried out on pad B. As mentioned earlier, water had been allowed to dribble from the pad water system pipes and was the source of the ice. The team poured 1,450 gallons of antifreeze into the water troughs at the base of the pad. Filled with 6,580 gallons of water, the troughs suppress the sound and vibration of rocket exhaust and dampen exhaust pressure, which is reflected back to the shuttle as it begins liftoff. This overpressure effect was discovered during the *Columbia* test flights. It created a potential for steering deflection in the main engine and booster motor nozzles at liftoff.

Stevenson and Davis made their first ice inspection at 1:30 A.M. on the 28th under pad floodlights. They were assisted by two technicians who were familiar with the water system. They found a sheet of ice covering about 3,000 square feet of the pad under the rotational service structure, ranging from a quarter inch to three inches thick.

On the mobile launch platform there was a sheet of ice one-eighth inch thick. A half-inch sheet of ice had formed in the overpressure water troughs. Icicles up to 2 feet long and three-fourths of an inch in diameter at the base hung from the fixed service structure between the 100- and 200-foot levels. One-eighth-inch-thick ice had formed on supports, electrical distribution panels, and valve panels.

"Upon return to the launch control center," Stevenson related, "we immediately held a meeting with our upper management members. . . . We concluded that the ice we had seen in the overpressure water troughs would not be acceptable for launch based on previous experience of ice coming out of the troughs."

The ice/frost team attempted to calculate the trajectories of icicles falling off the fixed service structure when loosened by engine vibration. Would they hit the heat shield tiles? While this analysis was being worked, the team made a second sortie to the pad at 6:54 A.M., accompanied by pad technicians. They worked for two hours to lift ice out of the water troughs with fish nets. Temperatures at the pad then ranged from 26.1° to 30.1°F, Stevenson said, while ice and unfrozen antifreeze-water mixtures in the troughs exhibited temperatures of 14° to 16°. Temperature was measured by an infrared pyrometer or Omegascope, a scanner with a pistol grip. It displayed the temperature of objects at which it was pointed.

On its return to launch control, the team met with launch managers headed by Aldrich. Trajectory calculations indicated that ice falling from the fixed service

Ice forming on the shuttle launch pad was a problem a year before the *Challenger* 51-L launch. In this photo taken January 22, 1985 before the launch of *Discovery* 51-C, a pad crew pours antifreeze into water of the sound suppression system below the solid rocket boosters at pad 39A. (NASA)

structure when the engines fired would hit the mobile launch platform deck about 20 feet away from the fixed service structure, away from the orbiter. It was concluded that the icicles on the service structure would not be a flight safety hazard, Stevenson said.

Aldrich asked for a third inspection, which is not normally done, Stevenson said. The team went back to the pad at 10:30 A.M. and returned 20 minutes before launch. Ice remaining on the west side of the mobile launch platform deck was removed. Temperature readings at the pad reached 34.8° to 36.2°F. More ice was fished out of the troughs. Stevenson said that he described the results of the inspection and ice clearance to Launch Director Gene Thomas. Testimony of the ice/

A year later, ice once more threatened a launch, this time of *Challenger* 51-L on pad 39B. This photo was taken during the ice inspection before the launch on January 28, 1986. (NASA)

frost team, which began February 26, continued for two days.

In response to questioning by Chairman Rogers, Stevenson said that Omegascope measurements had shown a considerable difference in readings between the two boosters. The left booster was recorded at 33° but the right booster was much colder, 19°. The lower temperature was the result of wind blowing off the supercold external tank, loaded with cryogenic propellant—liquid hydrogen and liquid oxygen—onto the right booster. But Stevenson did not report this variation in booster temperatures at the time, he said, because he believed along with the rest of the launch team that the shuttle would be launched well within launch commit criteria.

Although icing was extensive, it did not appear that

Icicles form on the pad structure adjacent to the right-hand solid rocket booster. (NASA)

the threat to the shuttle by the ice buildup on this launch was as significant as in previous freezes, Davis said. "But the temperatures were much colder than we have ever seen before," he said. "There was ice about one-eighth-inch thick for 30 feet up the side of the left booster, but none on the right, even though it was colder." Davis accounted for the ice on the left booster as the result of wind blowing water off the fixed service structure during the night.

Stevenson and Davis were questioned about the liklihood of a leak in the external propellant tank. None was observed, they said. Had there been a leak, it would have shown up on the infrared temperature scanner and also would have been visible as a cloud of condensation, they said.

Rockwell's Caveat

Testimony of the pad inspectors set the stage for the commission to press its inquiry into a second caveat

On the morning of the *Challenger* launch, this pad platform looked like an ice-skating rink. (NASA)

by prime contractor personnel concerning launch on the 28th. The contractor was Rockwell International, manufacturer of the orbiter. Its spokesman feared extensive damage to the orbiter heat shield, an intricate structure of 31,000 tiles. During the early flights of *Columbia*, when tiles had simply been shaken off by launch vibration, it had taken weeks to replace them. Ice threatened severe tile damage, exposing the orbiter

to refurbishing delay as well as to the peril of reentry heating. If a damaged vehicle escaped destruction in reentry, repairs to the shielding might well take months. Rockwell managers questioned the wisdom of taking such a risk.

On February 27, testimony of the ice/frost team was resumed. Chairman Rogers asked: "Mr. Davis, you said that weather conditions on the pad were the worst

A pad control fixture is frozen solid. Ice formed from water system drainage overnight to prevent pipes from bursting during the January freeze. (NASA)

of any previous launch. Is that correct? Or maybe it was Mr. Stevenson who said that.

"We probably both said it," said Davis. He added that ice on the launch pad was worse than he had seen it on any previous flight, although ice on the orbiter was not as extensive as during previous freezes. In addition to the 2-foot-long icicles on the fixed service structure and foot-long icicles on the rotational service structure at the 160-foot level, there was an array of small icicles at the tops of these structures. They would be heated by sunlight and be ready to fall at the time of liftoff, Davis said he had estimated.

On launch pad A, heavy canvas had been hung on the rotational service structure to protect the orbiter tiles from rain and hail during previous winter launches. This had not been done on pad B, which was being used for

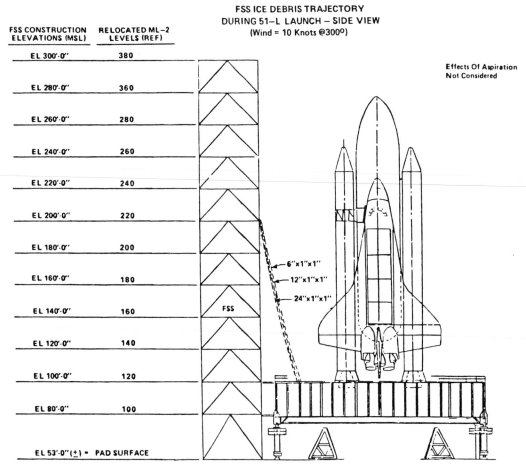

Calculations by the launch directorate showed that icicles hanging from the fixed service structure would not hit any part of the shuttle when shaken off by main engine and solid rocket motor firing. As noted in the sketch, the effects of ice particle impacts by aspiration (suction) were not considered. (*Commission Report,* vol. 3)

a shuttle launch for the first time. Consequently, as Stevenson explained, *Challenger* had no protection during the 37 days it stood there after it was rolled out from the Vehicle Assembly Building.[7]

The lack of protection on the pad puzzled Commissioner Neil Armstrong. He asked Kennedy Space Center Director Richard G. Smith whether the pad was designed to provide some protection to the shuttle in freezing weather. And if not, why not?

Smith: Neil, early in the program, it was recognized that we were not equipped, coming out of Apollo,

to handle freezing conditions in the water systems on the pad.

Armstrong: Why should that be? I mean, certainly any civil engineer knows that he has to handle the normal environment where he is building his building.

Smith: Well, the answer to that would be, yes, except [that in] the history of the past 10 or 15 years, freezing pipes in Florida in that area has been a very rare thing. The decision was made not to implement the insulation and so forth on the water systems but preclude rupture of the lines; that the risk [was so low] that the cost was not justified. We had the experience then in the January 1985 flight and we could not support the launch because of damage [from frozen water pipes] to the facility. And since that time we had implemented the plan

---

[7] The commission report said the potential for water in the joints was present because more than 7 inches of rain fell while *Challenger* was on the pad. Under freezing conditions, ice forming in the joints could interfere with secondary O-ring sealing, the report said.

by which we felt we could minimize the damage, could keep the firing systems and safety systems up and in a functional stage, fully recognizing that doing that we would have to bleed the systems, so we would have ice on the structure. We were not satisfied with that.

Stevenson showed pictures of launch pad B festooned with icicles like a tower in the arctic. One photograph in the ice/frost teams postlaunch inspection series showed *Challenger* lifting off.

Looking carefully at the photo, Commissioner Hotz commented, "it looks like there's a little puff of smoke off on the right-hand side there. Charlie [Stevenson], can you describe that for us?"

Stevenson: Yes, sir. That is the puff of smoke that has been released before to the press and to the world. That puff of smoke in that picture is, I believe, 100 inches or so tall and about 5 feet across.
Hotz: What is the time frame of that picture?
Stevenson: I believe it's about 2 seconds. I would have to go back and check the time.

The inspection team had photographed the launch in 130 sequential photos as part of their postlaunch observations. Stevenson said smoke had appeared for the first time on chart 24. In one photo, the cloud of smoke was 36 by 108 inches. In another, it was 72 by 130 inches.

Rogers asked Stevenson how he would interpret the puff of smoke and invited him to speculate if he wanted to.

"Engineers don't like to speculate," Stevenson said, "but based on our photo data, and we have analyzed all of the photos, we feel that it is a leak . . . we feel it is coming out of the— that the most likely spot is the joint between the aft booster and the aft segment."[8]

Rogers: And this is the right booster?
Stevenson: This is the right booster, yes.

[8] He probably meant the aft-center and aft segment joints of the right booster.

There was no sign of a leak on the external tank, Davis added. Rogers said: "So your speculation would be the same as Mr. Stevenson's?"

"Yes, sir," said Davis. He estimated that the ambient temperature at scheduled launch time (9:38 A.M.) was 28° to 30°F.

On questioning by Commissioner Ride, Davis said the smoke seemed to be coming from a spot inboard of the connection of the lower strut linking the right booster and the external tank.

"We'll say it is within a foot of the joint," said Stevenson, referring to the junction of the aft-center and aft segments of the booster.

Stevenson and Davis said that once "opinion developed" that ice would not interfere with the launch or damage the orbiter, "we had no problem with saying that it was okay to launch." They made this recommendation orally to Jesse Moore and Arnold Aldrich.

Rogers then asked whether anyone expressed doubt about the wisdom of flying in view of the cold weather and the ice.

"Yes, sir," responded Stevenson. "Rockwell, Rockwell expressed concerns. They did not express a strong opposition to the launch."

Rogers then called on Rockwell executives to tell their story. The first was Rocco Petrone, former NASA executive, who had left the agency after the Apollo project to join industry. At that time, he was president of the Space Transportation Systems Division, North American Space Operations, Rockwell International. Rogers asked Petrone to explain his concern about the weather.

Petrone testified that he had been at the Cape since January 24, but had flown back to the Rockwell plant at Downey, California after the scrub on January 27. He left his program manager, Robert Glaysher, vice president of orbiter operations; Martin Cioffoletti, vice president of Space Transportation Systems Integration; and Al Martin, site director in the company's launch support operations at the Kennedy Space Center. Petrone said he had received a call at Downey at 4 A.M. (PST: 7 A.M.

EST) on the 28th, telling him about ice concerns. He hurried to the mission support room at Downey to find out what was going on. He was told that the Rockwell people were expected to take part in a meeting at 9 A.M. (EST). Petrone said he was uneasy about the report from KSC.

"We had not launched in conditions of that nature," he said, "and we felt we had an unknown. I then called my program managers in Florida and said we could not recommend launching from what we could see. We think the tiles would be endangered. I said—let's make sure that NASA understands that Rockwell feels it is not safe to launch."

At the 9 A.M. mission management meeting on the 28th, Robert Glaysher said, he reported that "Rockwell cannot assure it is safe to fly." Al Martin said that the icicles could become debris when the engines fired and that their trajectories were unpredictable.

Rogers asked the Rockwell executives: "Did you convey to them in a way that they were able to understand that you were not approving the launch from your standpoint?"

Cioffoletti replied: "I felt that by telling them [the launch management team headed by Aldrich] we did not have a sufficient data base and could not analyze the trajectory of the ice—I felt he [Aldrich] understood that Rockwell was not giving a positive indication that we were for the launch."

Glaysher summarized: "And so we felt that we had communicated Rockwell's position that we felt it was unsafe to fly."

Commissioner Ride: Had Rockwell ever taken that position before on previous launches when the launch had occurred?
Glaysher: No. This was the first time where we had been in a position where we really had no data base from which to make a judgment, and this was the first time that Rockwell has taken an unsafe-to-fly position.

Rogers asked Petrone: "And your recommendation, now you say, it was unsafe to fly?"

"Correct, sir," said Petrone. He then related that he had received a call from Al Martin following the morning meeting saying that it looked as if NASA was going ahead to launch. "We knew that was within a few minutes of the loading of the crew."

Armstrong: Now clearly when they resumed the count, you knew that your recommendation essentially had been either considered and overruled or dispositioned in some other way.
Petrone: That's right, sir.

Armstrong asked if Petrone had expressed an opinion about that.

Petrone: Mr. Armstrong, I felt we had expressed our opinion to the proper level on the proper occasion of the meeting that had been set up for it. I felt that I had done all I could do.
Ride: Did it surprise you that NASA picked up the count?
Petrone: I was disappointed that they did, yes.

The testimony showed once more that a reversal had occurred in the traditional launch decision process. Instead of being required to prove that all systems were go, a second prime contractor had been put in the position of proving they were not. It was as if NASA managers were either deaf to or unimpressed by contractor caveats about weather conditions.

Oversight

As the hearing continued on February 27, Arnold Aldrich and Jesse Moore took up the recital of what the managers knew or did not know the morning of the launch. It was evident that what they did not know was what one commissioner had described as the "tenderness" of the O-ring seals and the effect of temperature on them.

Aldrich, who chaired the mission management team, described the outlook after the 27th when the launch was scrubbed. With a forecast of subsiding winds and clearing on the 28th, he said, he did not expect a problem with the weather. Cold was the only issue. When the NASA and Department of Defense Members of the mission management team met in the afternoon, "we asked each of the projects—the Kennedy launch and landing project, the orbiter, the external tank, the solid rocket boosters, and the main engines—how they felt about launching with the temperature as predicted. The one major concern was the cold temperature."

Associate Administrator Moore was present and agreed to Aldrich's recommendation to proceed with filling the external tank with orbiter engines propellant for launch the next morning. However, Aldrich said, he remained concerned about ice on the pad.

A measure of the team's lack of awareness about the tenderness of the booster seals was this focus of the members on the ice, not on the low temperature per se and its effect on flight safety. In terms of the probable cause of the accident, the managers' weather concerns overlooked or miscalculated temperature effect on the booster seals. Thus a crucial safety threat was overlooked as the result of human failure in communication. This was the flaw in the launch decision process perceived by the commission, and the evidence showed that it was fatal.

Hindsight suggests comparison with the oversight that contributed to the Apollo 1 fire of 1967. Launch managers then failed to take into account a high probability that flame-resistant plastics in the Apollo command module would burn explosively in a high-pressure atmosphere of pure oxygen when ignited by an electrical spark. Poison gas from the burning plastics quickly suffocated the flight crew, who were running a prelaunch test of the module on the launch pad.

If these two tragedies demonstrated anything, it was that oversight is the ultimate human failure in technological development; realizing it is a bitter learning experience. At this point in the *Challenger* accident investigation, the human failure was being defined as contributing to the mechanical failure.

Aldrich told the commission that on the night before the launch, he received a call at the motel from Mulloy and Reinartz of Marshall expressing their con-

cern for the booster recovery ships. The ships were reported as beset by high seas.

Chairman Rogers interposed: "At that time did they tell you that there had been serious concerns expressed by Thiokol and Thiokol engineers; that they had a long teleconference on the subject; and that first Thiokol had recommended against launch and secondly management in the person of Mr. Kilminster had changed its mind, and Thiokol then had decided to recommend launch? Did you know of that sequence at all?"

"None of that was discussed," said Aldrich. "And I did not know until after the 51-L launch that there had been such a meeting."

He went on to relate that at 3 A.M. on the 28th he received a call from his deputy, Richard Kohrs, saying that the countdown was proceeding but that they were one hour behind schedule because of an electronic problem in ground equipment. There was also an ice problem, Kohrs reported. Everything in the countdown would be running one hour later than the schedule.

When he arrived at the firing room at 4:30 A.M., Aldrich said, he was told that the offshore winds had subsided, and that the recovery ships would be able to reach their stations in time to recover the boosters after the launch. This reduced the weather concern to the possible effect of pad ice on the orbiter at launch.

Aldrich then described the scene before the launch, depicting for the commission and through it, the public the transcontinental operations involved in launching the shuttle.

Adjacent to the firing room where the launch was controlled was an alternate firing room where engineering managers for the external tank, the orbiter, the main engines, and the boosters were stationed, along with key technicians who analyzed problems as failure modes developed. Supporting the orbiter team was an engineering team in Houston in the mission evaluation room of the Johnson Space Center. They were in constant communication during the countdown.

In California, there was a mission support team of engineers at the Rockwell facility at Downey. They had data that would help mission managers deal with engineering problems in flight. In Alabama, there was an external tank support facility at the Marshall Center. It was part of the Huntsville Operations Support Center where engineers familiar with the tank, the boosters, and the main engines were in contact with Kennedy launch control. Reports by the ice/frost team had been relayed to the Huntsville center.

Aldrich related that he, Moore, Reinartz, and Lucas sat in the operations support room, a glassed-in cubicle overlooking the firing room. About 8:30 A.M. the ice team reports and engineering assessment of them indicated that ice conditions on the pad were looking more favorable for the launch. Aldrich said he called a mission management team meeting to review the situation. It was held in a conference room on the floor above the control center at 9 A.M.

At this time, Aldrich said, he sought opinions on the severity of the ice threat. The response of all parties in NASA, he said, was a recommendation to proceed with the launch. He polled Glaysher and Cioffoletti of Rockwell. Glaysher responded that Rockwell would not give an unqualified "go" because the ice posed an unquantifiable risk, Aldrich said. He added: "Mr. Glaysher did not insist that we not launch, however."

"A Hell of a Problem"

At the conclusion of the partial management review that morning, Aldrich said "I felt reasonably confident the launch should proceed." However, he asked for an additional inspection by the ice/frost team as close to the launch as possible. Thus, the team went out a third time. Aldrich said he then returned to the operations support room to review the results of the management meeting with Moore and Philip Culbertson, NASA general manager.

Aldrich said he summarized the situation. He mentioned the qualified position taken by Rockwell. He recommended that the launch proceed unless the ice team identified a significant change in pad conditions after their third visit. They did not report a significant change.

Rogers asked if there had ever been an instance when a prime contractor voiced objections to a launch and the launch directorate went ahead with it notwithstanding the objections. Aldrich replied that he interpreted Rockwell's input as a "concern, not an objection to the launch."

Said Rogers: "If the decision-making process is such that the prime contractor thinks he objected and testified under oath that they took a position it was unsafe to launch, and you say that it was not your understanding, that shows us serious deficiencies in the process."

Aldrich argued that Rockwell had taken positions before, similar to the one it took on 51-L. When Rogers asked for examples, Aldrich could recall only one instance, where the contractor was overly concerned about insulation coming off the external tank and hitting the orbiter.

"It was a very familiar kind of report and discussion to me," he said.

Rogers then read a statement from an executive session in which the Rockwell caveat was discussed. Dr. Sally Ride had commented: "Well, I guess the question is whether at the end of the meeting, Rockwell was saying, 'we don't want to launch.'" To this, Rogers had added: "And I said, 'that is exactly it. If Rockwell comes up in a public session and says we advised NASA not to launch and they went ahead anyway, then we have got a hell of a problem.' I guess I shouldn't have said that."

Aldrich then explained his role in making launch decisions. "Mr. Chairman," he said, "I would like to describe for you at some other point the kinds of decisions I have had to make during the last six launches in terms of marginal conditions and weather threats to the orbiter in flight and how I believe this kind of decision couples rather closely to the kind of decisions that I have to make for launching with rain clouds in the area, launching with the potential for low ceilings or cross-

Chairman William P. Rogers commenting on the ice situation: "If Rockwell comes up in a public session and says we advised NASA not to launch and they went ahead anyway, then we have got a hell of a problem. I guess I shouldn't have said that." (NASA)

winds at the return to launch landing site and overseas . . . what I did was my responsibility and I executed it in the same manner as I have on other flights for other conditions. They were also a threat and also marginally acceptable or unacceptable based upon specific assessment of the conditions at the time."

Rogers said that he recognized the difficulty of Aldrich's position, "and we respect it and respect you." He went on: "Obviously your interpretation of what

Rockwell said this morning is somewhat different than Rockwell's interpretation of what they said."

Aldrich answered: "If Rockwell had told me that they were no-go, I would have reported to you in the manner of George Hardy. . . . I would not have overruled a no-go decision from the Rockwell team."[9]

[9] As mentioned earlier, Hardy had testified that he would not have approved the launch against a no-go recommendation from Thiokol management.

In response to a question from Commissioner Hotz, Aldrich reiterated that he was not aware of temperature concerns about booster seals. This question kept bubbling up from time to time. It seemed difficult for commissioners to believe that not a word of the three-hour argument between Marshall managers and Thiokol engineers about the low-temperature effect on the seals had passed up the line from Level 3 to Levels 2 or 1. One might have expected normal gossip that is endemic at the space centers to bridge the communications gap.

Commissioner Armstrong asked if the launch window could have been reset from morning to afternoon when the weather was expected to be warmer.

"It could have been if planned earlier in the day," said Aldrich. "We had lights at Dakar and that was go,[10] but we could not go later in the day because of the schedule we had set the launch crew and the flight crew on" (the sleep-wake schedule).

Also, Aldrich said, there was a time limit for the crew to remain in the orbiter in launch position, and it ended at 12:30 P.M. The crew's sleep-wake cycle would enable the members to do a full day's work in orbit with a morning launch. It would have been feasible to change to an afternoon launch if that had been prepared several days earlier, he said.

The View from the Top

Following a lunch break, the commission resumed the hearing in the afternoon of February 27 by calling William R. Lucas, director of the Marshall Space Flight Center, as a witness. Lucas had been with Marshall in engineering and management positions more than 34 years. He had joined the Army rocket research and development project at the Redstone Arsenal (from which the Marshall center evolved) in 1952 and the Army Ballistic Missile Agency in 1956. He was a lead mate-

rials engineer in the development of the nose cone for the Jupiter intermediate ballistic missile. The nose cone was the invention that made possible the U.S. ICBM and manned space flight by providing the shielding necessary for a vehicle to reenter the atmosphere from space.

A native of Newbern, Tennessee, Lucas was graduated with a bachelor of science degree in chemistry from Memphis State University and earned a Ph.D. in metallurgy from Vanderbilt University in Nashville. He emerged from the Wernher von Braun era at Marshall as one of NASA's top engineer-scientists. He had received NASA medals for exceptional scientific achievement and for exceptional services in Project Apollo, along with NASA's Distinguished Service Medal. In 1980, President Carter conferred on Lucas the rank of distinguished executive.

Now Lucas, whose experience in rocketry spanned the space age in America, was being called upon to explain what he knew about the booster seal problem and what he did about it.

Chairman Rogers asked the Marshall director when he first learned of the concern of the Thiokol engineers about the low-temperature effect on the *Challenger* booster seals. It was in the early evening of January 27, about 7 P.M., Lucas said. He was talking to James E. Kingsbury, Marshall science and engineering director, at a space coast motel, when Reinartz and Mulloy entered the room. They reported that some members of Thiokol had raised a concern about the performance of the boosters at the low temperatures expected for the next day. Lucas said he asked them to keep him informed of the outcome of the planned teleconference. He said he heard nothing further about it until he went to the launch control center at 5 A.M. the next day and asked Reinartz and Mulloy how the concern was handled.

Lucas said that he had been aware of the case joint seal problems since the beginning of the shuttle program. Evidence of minor seal erosion or blow-by, first appearing on the second flight of *Columbia*, had been

---

[10] Dakar, Senegal was the first transatlantic abort emergency landing site. In the event of a late afternoon launch, it would be dark there.

"considered and dispositioned on each and every flight readiness review," he said. "So I am familiar with that part and never considered the seals, however, a safety-of-flight issue."

A Reasonable Risk

However, he said, when secondary O-ring erosion appeared in April 1985, "we did become concerned and began to accelerate ways of improving, increasing the margin on that seal." Lucas was aware that the seals were a criticality 1 item in terms of the program, but he had not considered them a threat to flight safety.

Rogers: Were you familiar with the solid rocket booster critical items lists that were signed December 17, 1982?

Lucas: Yes, sir.

Rogers: And didn't that indicate to you that there was a serious problem of flight safety?

Lucas: It indicated to me that if we had a failure of the primary O-ring, after rotation of the joint, and the secondary O-ring had opened up and didn't seal, that there would be a problem. But never in our experience that I'm aware of had that happened.

Rogers: Well, I was just commenting on your testimony you never considered it as a matter of flight safety. It seems to me that the critical items list and the waiver goes directly to that. It says 'actual loss, loss of mission, vehicle and crew due to metal erosion, burn-through and probable case burst, resulting in fire and deflagration.' I mean I don't see how you could say that didn't involve flight safety.

Lucas: Well, if it happened, it would involve flight safety. My conception was that it was a reasonable risk to take in criticality 1.

Even after April 1985, Lucas continued, he didn't think the problem was serious enough to ground the fleet, although it seemed more serious than it had earlier. Lucas, however, had not heard of alarms that had been sounded by Thiokol engineers in the summer of 1985 and the night before the launch.

Rogers: And if you had heard those alarms you would have been concerned about flight safety, I presume, wouldn't you?

Lucas: If I had heard the alarms that have been expressed in this room this week before the flight, I certainly would have been concerned, yes, sir. That is right.

Rogers: And you heard Mr. Reinartz say he didn't think he had to notify you, or did he notify you?

Lucas: He told me . . . when I went into the control room that an issue had been resolved, that there were some people at Thiokol who had a concern about the weather; that had been discussed very thoroughly by the Thiokol people and the MSFC [Marshall] people and it had been concluded that there was no problem, and that he had a recommendation by Thiokol to launch and our most knowledgeable people and engineering talent agreed with that. So from my perspective I didn't see it as an issue.

Rogers: And if you had known that Thiokol engineers almost to a man opposed the flight would that have changed your view?

Lucas: I'm certain that it would.

Rogers: So your testimony is the same as that of Mr. Hardy. Had he known, he could not have recommended the flight be launched that day.

Lucas: I didn't make a recommendation one way or the other. But had I known that, I would have interposed an objection, yes.

Rogers: I gather you didn't tell Mr. Aldrich or Mr. Moore what Mr. Reinartz had told you?

Lucas: No, sir. That is not the reporting channel. Mr. Reinartz reports directly to Mr. Aldrich. In a sense, Mr. Reinartz informs me as the institutional manager of the progress that he is making in implementing his program, but I never on any occasion reported to Mr. Aldrich.

Rogers: And you had subsequent conversations with Mr. Moore and Mr. Aldrich prior to the flight, and you never mentioned what Mr. Reinartz had told you?

Lucas: I did not mention what Mr. Reinartz told me because Mr. Reinartz had indicated to me there was not an issue, that we had a unanimous position between Thiokol and Marshall Space Flight Center and there was no issue in his judgment nor in mine as he explained it to me.

The Marshall director went on to say that when he asked about the resolution of the teleconference, Mulloy or Reinartz—he thought it was Mulloy—showed him the telefax from Kilminster recommending the launch.

Because of that piece of paper, Rogers commented, it could be argued that Reinartz lived up to the book, "but there was no application as far as you can tell of common sense."

The chairman continued: "I mean this obviously was a very serious matter, and by insisting that this piece of paper was giving everybody the blessing of Thiokol, the fact was that the top people who made the decision never knew about what happened in this long telecon and didn't know that the decision to launch, recommending a launch, was made really by just a couple of people, Mr. Kilminster and Mr. Mason and maybe one other. The fourth gentleman, Mr. Lund, said he was in a position where he couldn't prove that it was not safe and therefore he put on his management hat and changed his mind.

"So this piece of paper which really resulted in the launch was made by just a couple of people and apparently none of you gentlemen knew about it."

Lucas said he did not know that it was made just by a couple of people. He said he recognized Kilminster as the senior Thiokol individual for the space booster program.

Rogers pulled back a little. He said: "Well, maybe I am not being conservative enough. It was Mr. Kilminster, Mr. Mason, Mr. Wiggins,[11] and I think a fourth one was Mr. Lund. But Mr. Lund said—well, I will read you what he said because I think it illustrates the whole problem. Here's his testimony: ' . . . we got ourselves into the thought process that we were trying to find some way to prove to them [NASA] that it wouldn't

[11]C. G. Wiggins, vice president and general manager, Morton Thiokol space division.

work. We were unable to do that. We couldn't prove absolutely that it wouldn't work. This is the kind of boat we got ourselves into that evening.'"

Commissioner Hotz asked: "Dr. Lucas, did you know that Mr. Kilminster earlier had formally recommended against launch and then had reversed his position?"

Lucas said he had heard that Kilminster had said he couldn't go against his engineering and called for a caucus to discuss it. "It is not unusual in our system for one or more engineers to raise a concern and then have those concerns discussed and threshed out."

"Mr. Lucas, that's not the testimony, though," said Rogers. "The testimony was they had a long teleconference and Thiokol made a formal recommendation against the launch. It wasn't just casual conversation. They had a chart there. The chart said we recommend against launch. . . . They made a formal representation, no launch, and then they had a long, off-the-record or off-the-telephone conference caucus and it turned out that Mason, Wiggins, and Kilminster supported this document [the recommendation to launch] and all the engineers were against it, and Lund said, well, I'm chicken, I have to go along. I can't figure out a way to prove it's not safe."

Lucas said he had heard that in the testimony that week, but did not know it at the time of the teleconference.

"Well, in any event," said Rogers, "at no time did you pass on the information you had, even though it was sketchy, to either Mr. Moore or Mr. Aldrich."

Lucas agreed he had not, but reiterated that the project channel was from Reinartz to Aldrich. Still, from what Reinartz and Mulloy had told him, he did not consider the events of the night of the 27th an issue.

What They Didn't Know and When They Didn't Know It

The commission next heard from NASA executives who were involved in the launch decision-making process but who had not been privy to the no-go—go reversal. They were Associate Administrator Moore, Shuttle Manager Aldrich, Kennedy Space Center Director Smith, and Launch Director Thomas.

Rogers asked if any of them knew about the Thiokol objection to the launch before launch time. Each replied he did not know.

Rogers: So the four—certainly four of the key people who made the decision about the launch were not aware of the history that we have been unfolding here before the commission?
Moore: That is correct.

Rogers than asked Moore and Aldrich if they knew the origin of the change in the primary O-ring seal in the field joints from criticality 1-R to 1. Moore said the change was made before he joined the shuttle program in 1983, and he did not have the background on it.

Rogers asked Aldrich if he knew. ''No, sir,'' he replied. ''I have not studied this critical items list item on the SRB in detail.''

''Let me try another tack,'' said Rogers. ''After you make this change and you issue this piece of paper and it circulates and everybody has to be more careful, what in fact happens? Is there a different treatment, when you get ready to launch, of these items, or is this just paperwork?''

Aldrich then explained in detail the purpose and theory of the critical items list and the manner in which it is established. He summarized by saying that criticality 1 and 1-R items deal with the safety of vehicle. After being formulated at lower levels, they are submitted to Level 1 in the NASA hierarchy at headquarters for critique, review, and approval.

Rogers: So if I understand what you are saying, that meant that you and Mr. Moore would be very conscious of this problem because it was a criticality 1 item. Is that right?
Aldrich: I would be very conscious of it if it had been brought to my attention.
Rogers: Now what requirement was there to bring it to your attention?

Aldrich: The requirement on the project is that these items be fully documented and fully approved, and this one—
Rogers: But you know, that doesn't really mean much to me, Mr. Aldrich. I mean, the people that have testified here say they are Level 3; they didn't have to do anything except get the pieces of paper and handle it at Level 3, and I am asking, did this require something else, that because of the criticality nature of this and because it says that the loss of mission and vehicle and crew will result from it, did that put added burden on the people in the decision-making process to notify you people at the top about it?
Aldrich: The people at the top of the program were notified in 1983—and please let me finish. As you mentioned, there are a large number of these criticality 1 items and they are not each individually and uniquely reviewed for each launch. Changes to those criticalities in terms of the rationale, the understanding of how the system performs, and any failures that occurred are highlighted, and it is the responsibility of the projects, the contractors to notify their project office at NASA and up the chain of changes to performance of the equipment or to the rationale that applies to the category of these criticality 1 items.
Rogers: Did that mean, then, that in 51-C you were all notified about what happened on that and the fact that it was the coldest day and the most serious blow-by? Did you know about that?
Aldrich: Well, the criticality 1 item was not brought up for attention to my level: that there was a change either in the engineering assessment of how the system performed or in the flight experience that would change the rationale that is documented here. . . .
Rogers: Let's go back to my question. Did you know about 51-C and the fact that it was the coldest launch and you had the most trouble with this seal?
Aldrich: I knew that 51-C was launched under cold conditions. I knew—I did not know at the time that 51-C had had the most blow-by or the most erosion, if that in fact was true.
Rogers: None of you did, I gather.

Moore: No, sir. Neither did I. I knew that 51-C was a cold launch because I remember we scrubbed the [scheduled] day of the launch because of excess ice on the external tank, and we were worried about the thermal protection system [orbiter heat shield]. But I did not recall any correlation between temperature and the erosion experience that we had seen on 51-C.

At this point, Major General Donald Kutyna, commissioner, asked how the briefing that Marshall and Thiokol had presented on O-ring problems in August 1985 had been dispositioned at headquarters.

Moore replied that the review had been a result of the memorandum from Irving Davids, headquarters engineer, describing instances of erosion and blow-by on previous flights. After the 51-B flight of April 29, 1985, when secondary seal erosion in the nozzle joint was experienced for the first time, "we got a little bit more concerned about the erosion problem at that time."

Because of the press of other duties, Moore said, he had not attended the Marshall-Thiokol briefing on August 19. It was attended by L. Michael Weeks, his deputy; David L. Winterhalter, head of the propulsion division at NASA headquarters; and William Hamby, deputy director of shuttle program integration. Marshall attendees were Mulloy and Robert Schwinghamer, director of the materials laboratory. Six Thiokol representatives were there, including McDonald, Mason, Killminster, and Wiggins.

Weeks had reported to Moore later that he had listened to the briefing and believed that the data basically supported a policy of continuing to fly. "He did not think it was an issue that we ought to ground the fleet," Moore said.

Commissioner Arthur B. C. Walker, Jr. asked that since the change in O-ring criticality from 1-R to 1 had occurred in December 1982, why did a program to restore redundancy to the joint seals not begin until April or so of 1985?

Moore explained that planning for tests and analysis had been going on since 1984.

"In view of that," said Rogers, "It didn't occur to any of you to ask the question about the weather and the seal problem prior to launch?"

Moore said it was his assumption that the hardware was qualified to operate between ambient temperatures of 31° and 99°F. Had he not heard the O-ring problems and the possibility that the seal was harder when colder? asked Rogers.

"I had not heard any of that information that you are commenting on, Mr. Chairman," Moore replied.

KSC Director Smith said that he had not heard that information either.

Aldrich said that the solid rocket booster project had been committed intentionally to the 31°–99° temperature range. "If there was a concern with that, they would be required to submit additional launch criteria to us that said for a given solid rocket booster system there was a different temperature range—and we would honor that constraint."

"Did any submit such amendments?" asked Commissioner Armstrong.

"No, sir," said Aldrich. "Not to my knowledge."

In reply to a question from Commissioner Covert, Aldrich said that a separate certification existed for the bulk propellant in the motor, with temperature limits of 40° to 90°F. This range was relevant principally to conditions of shipping the booster segments by rail or storing them, but the bulk propellant was not readily affected by ambient temperature changes on the launch pad.

Rogers commented on the paperwork in connection with the boosters. The trouble with it, he said, was that "you eliminate good judgment and common sense." He said he could not imagine why some of the people who knew about the seriousness of the seal problem did not pick up the telephone themselves and call headquarters.

Looking back on the amount of discussion that had gone on about the O-ring response to temperature, said Moore, he would have thought the problem would have been brought forward to Level 2 (Aldrich's level), "if you want my honest opinion."

The executive director of the commission, Alton G.

Keel, Jr., asked Moore if he knew that Rockwell had warned that it was not safe to fly because of ice. Moore said he had some indication of Rockwell's attitude from a report by Aldrich, but that Aldrich did not indicate that it was a flight safety concern. The thought did not cross his mind that Rockwell was saying no-go, Moore said.

"No, they did not make that recommendation," said Kennedy Space Center Director Smith. "They did not concur, and I have heard people non-concur."

"I believe," said Rogers, "that I am speaking for the whole commission when I say we think it [the decision-making process] was flawed."

The final witness of the series of hearings of February 25–27 was Ben Powers, an engineer in the propulsion laboratory at Marshall for 20 years.

Chairman Rogers referred to testimony that George Hardy, deputy director of science and engineering at Marshall, had said he was appalled by the no-go recommendation of the Thiokol engineers. He then asked Powers what he thought about that recommendation. Was he, too, appalled?

"Sir," said Powers, "I fully supported the Thiokol engineering position and was in agreement with it."

Had Powers made that known to Hardy? asked Rogers. Powers said that he reported his concurrence with the Thiokol engineering position to his superior, the deputy director of the laboratory, John McCarty, and to the chief engineer, Jim Smith. "This would be the typical thing that we would do," he said.

"Were there others that agreed with you?" asked Rogers. Powers said he could not identify anyone, but on further interrogation said: "Some of the engineering people have mentioned that they, too, were concerned, primarily with the temperature effect on the O-ring resilience, the spring-back ability of the O-ring."

Rogers: Was there anybody who agreed with Mr. Hardy or Mr. Mulloy, as far as you remember, on that telecon?
Powers: There was no dissent with Mr. Hardy to my knowledge other than the discussion that I had. I was the only dissenting engineer.
Rogers: But the others remained quiet, I assume?
Powers: Yes, sir.

7. Salvage

The physical evidence necessary to confirm the cause of the *Challenger* accident, as indicated by telemetry data and film, lay on the bottom of the Atlantic Ocean. It had to be recovered and analyzed before the cause and effects of the accident could be understood. That had to be done before the shuttle would fly again.

Retrieving sufficient wreckage to satisfy these requirements was to take three months and develop, as noted previously, into the most extensive salvage operation in world maritime history. The search for debris covered the ocean off the east coast from Cape Canaveral, Florida to Cape Hatteras, North Carolina and ranged 90 miles out to sea. Underwater salvage was carried out by 15 surface ships, scores of divers, and five submarines including the Navy's nuclear-powered research submarine, NR-1. The operation began February 6 after tons of floating debris were picked up by NASA booster retrieval ships and Coast Guard vessels, with Coast Guard, Air Force, and Navy helicopters and fixed-wing aircraft flying search missions.

The main salvage object was the right-hand solid rocket booster, or what remained of it after it was detonated by the range safety officer. If the scenario of a leak of hot gas in the joint between the aft-center and aft segments of the boosters was accurate, there would be a hole in the rocket casing where the 5,600°F motor exhaust had burned through and gone on to pierce the external tank.

Finding the key evidence amid tons of wreckage strewn over hundreds of square miles of sea bottom was compared in a Navy report to seeking a dime dropped in several feet of muddy water in a football field where the searcher had to crawl about on hands and knees feeling for the coin. The strategy was to plot the trajectories of major pieces of debris from radar traces and establish impact areas; then locate the pieces by side-scan sonar towed by ships crisscrossing these areas; then obtain video images of the debris from manned and remotely controlled submarines. Divers or submarines with remotely controlled manipulators would then attach cables to the significant pieces so that surface vessels could haul them up on deck and move them to Port Canaveral.

Following the onset of the fireball at 73.2 seconds after liftoff, the boosters were detonated at 110.3 seconds, and the shuttle broke up at 113 seconds. Still ascending, most of the pieces reached maximum altitude of 122,400 feet at 146.3 seconds before beginning to fall into the ocean. Impact was established at 286 seconds after liftoff, or 11:42:46 A.M., EST, according to radar data.

At about that time, NASA's booster recovery ships, *Liberty Star* and *Freedom Star*, waiting on station at the planned impact area of a normal launch, were notified of the accident by radio. Their crews had not seen it because of clouds. Given the approximate location of the impact area, the vessels moved shoreward at top speed, about 15 knots. En route, they began to pick up floating debris.

As debris kept falling, an H-3 helicopter (*Jolly 1*) flew out to observe it and report when it would be safe for recovery vessels to enter the impact area. The Air

The solid rocket booster recovery ship, *Freedom Star*. (NASA)

Force missile range safety officer estimated that debris would continue to rain down for 55 minutes. *Jolly 1* took up a station away from the impact zone and maintained a stream of reports on the falling debris. At 12:37 P.M., the range safety officer broadcast a clearance for aircraft to enter the impact zone. A second H-3 helicopter *(Jolly 2)* signaled NASA's Support Operations Center that it was airborne and flying to the zone. Then a C-130 aircraft and HH-3 helicopter took off from the Clearwater, Florida Coast Guard station and flew across the peninsula to the impact area.

Meanwhile, at the Pentagon, the Joint Chiefs of Staff set up a shuttle response call that made the search and rescue resources of the armed forces available for recovery operations. From the Kennedy Space Center, a NASA official called the Harbor Branch Foundation at Fort Pierce, Florida to determine whether their research submarines and tenders were available for salvage work. Harbor Branch, a private ocean research institution, was destined to play a major role in finding the "smoking gun," the right booster joint from which hot gas erupted, then pierced the external tank and ignited its hydrogen to cause the fireball. From the Navy's Office of the Supervisor of Salvage, Captain Charles Bartholomew sent a representative to Cape Canaveral to make an assessment of requirements for the salvage operation.

The center of the principal impact area appeared

Debris from fuselage of the orbiter and external tank laid out in storage on Cape Canaveral. (NASA)

to be about 20 miles out to sea on an easterly bearing from the Cape. By the afternoon of January 28, 3 Coast Guard Cutters were approaching the area to assist the NASA ships in picking up floating debris. By January 30, 12 aircraft of the Air Force, Navy, and Coast Guard and 10 NASA, Navy, and Coast Guard ships were searching an ocean area of 1,200 square nautical miles. Hundreds of pounds of floating debris—principally from the light-weight structure of the external tank—were picked up and brought into Port Canaveral by these vessels.

NASA set up three depots for storing and recon-structing the wreckage. One was the Logistics Building and an adjacent area at the Kennedy Space Center, where orbiter debris was placed for reassembly. Another was Hangar O on Cape Canaveral, and the third was the Explosive Ordnance Disposal impoundment on the Cape where unburned solid propellant was stored and burned.

The National Transportation Safety Board was char-tered by NASA to perform a structural evaluation of recovered debris to determine the mode of breakup. As part of the initial NASA Mishap Board, later the Data and Design Analysis Task Force, a search, salvage, and

Shuttle debris for reconstruction of accident effects by the National Transportation Safety Board and the Data and Design Analysis Task Force is arranged on the floor of the Logistics Building at Cape Canaveral. (NASA)

reconstruction team was formed. It actually managed the salvage operation in conjunction with the Navy. The lead officer was Colonel Edward A. O'Connor, Jr., director of operations, 6555th Aerospace Test Group, at Patrick Air Force Base.

By January 31, the search area had been extended 90 miles offshore. The Coast Guard buoy tender, *Sweetgum,* brought in a recovered booster nose cone that night. It contained a parachute and four booster separation motors that under normal flight conditions would push the booster away from the external tank after burnout.

On February 1, the search area was extended northward to Savannah, Georgia. The broad search area ranged from Cape Canaveral to Savannah and 90 miles out to sea. Between Ponce de Leon and St. Augustine (about 60 miles) 7 ships and 14 aircraft patrolled. A dozen Coast Guard auxiliary ships were assigned to cover a near shore area, about three miles out from the surf, between the Cape and Jacksonville.

As the search moved farther offshore, NASA's *Liberty Star* and other ships encountered currents of the Gulf Stream. The crew reported difficulties deploying the sidescan sonar. The currents also interfered with the

Wreckage picked up from the Atlantic Ocean surface by the Coast Guard and Navy is unloaded at the Trident Basin, Cape Canaveral from the Coast Guard Cutter *Dallas* on the evening of January 30. (NASA)

reconnaissance of the unmanned small submersibles by moving them away from their targets.

Mowing the Lawn

The first phase of the salvage operation was the collection and analysis of all visual sightings and data to plot impacts on charts, defining the area where debris might be found. Using computers, the salvage team made mathematical models of the debris trajectories, from which fields of debris were targeted.

Ships were deployed to tow the sidescan sonars through the general area and locate debris. The locations were then charted. The acoustic signal reflections could not discriminate between pieces of shuttle wreck-

More floating wreckage brought in by the *Dallas* was unloaded through the night. (NASA)

age and refuse from passing ships, such as oil drums, miscellaneous hardware, even a kitchen sink. Identification was another operation.

The signals were generated in a 4-inch cylinder, 4 feet long, weighing about 50 pounds. These sonars were towed behind a ship about 10 to 50 feet above the ocean floor, emitting a constant signal stream. As the signals struck objects, they were reflected back to the receiving apparatus, a hydrophone.

The strength of the signal was proportional to the size, shape, and material of the objects. All contacts were finally numbered sequentially and plotted on a navigation chart in degrees of latitude and longitude.

Search tactics were to start at one end of a charted search area and map a swath of ocean floor with sonar. When one swath was completed, an adjacent swath was covered. The Navy operators compared the process to mowing a lawn. To avoid leaving gaps, progress was deliberate and slow, covering only a few square miles a day where debris was concentrated.

The efficiency of the acoustic detection process was affected by currents, sea turbulence, and the volume of debris in several places. Once an object was located, it was photographed by the manned or remotely controlled submersibles using high-resolution video cameras. Sometimes, NASA or contractor engineers aboard ship could identify the video images as they appeared on monitors in real time; otherwise, they were identified from tapes. Not every piece of debris was recovered. Only those pieces believed to be significant in reconstructing the booster failure and breakup of the orbiter were retrieved. They constituted about 30 percent of the entire shuttle by the end of April.

Retrieval of floating debris was accomplished first. It progressed rapidly and was terminated February 7. By that time, most of the floating pieces had either been picked up or had slowly sunk. In the first 24 hours of the salvage effort, about 600 pounds of material were picked up. The largest measured 15 by 15 feet, a sheet of aluminum from the external tank. In two days, about 1,600 pounds of floating debris had been picked up, the largest 30 by 5 feet from the tank.

Portions of *Challenger's* fuselage were located between 20 and 40 miles off Daytona Beach, the largest 25 feet long. The nose cones of both boosters with main parachutes packed inside were located nearer the coast.

On January 31, NASA's manager of aviation safety, Lieutenant Colonel Bill Comer, U.S. Army, arrived from headquarters representing the chief engineer. Working with Air Force counterparts, the Army engineer devised a system for assessing and storing the recovered debris. Reconstruction of parts of the shuttle was started by technical experts from the National Transportation Safety Board, who proceeded in a manner similar to that of reconstructing a crashed aircraft.

The February Catch

By the end of the sixth day of salvaging, more than 22 tons of debris had been picked up from the surface by Coast Guard, Navy, Air Force, and NASA vessels. A total of 29 ships of all types had searched 30,000 square nautical miles of ocean. Thirteen aircraft had flown 260 hours in search patterns covering 60,000 square miles.

As debris drifted with winds and currents and dispersed, the search rapidly enlarged. Helicopters ranged 60 miles north of the Cape to Flagler Beach and fixed-wing aircraft continued to reconnoiter northward 180 miles to Charleston, South Carolina. Fishing vessels occasionally came across pieces of debris and picked them up when it was possible. One piece was 25 by 4 feet.

The salvage team worked out definite priorities for debris recovery. The highest priority was placed on the right booster; next, the left booster; then the crew module, the orbiter wing, and components of the tracking and data relay satellite *Challenger* was carrying in its cargo bay.

On February 4, 4 ships and 6 aircraft were searching 12,500 square nautical miles of sea from New Smyrna Beach to Charleston. NASA's *Liberty Star* was towing a sonar tube 40 miles offshore in water 1,100 feet deep when reflections indicated a possible booster on the bottom. The contact was assigned a priority.

The next day, the Coast Guard extended the search area to Cape Hatteras, the farthest north of the search. The salvage team leader, Colonel O'Connor, reported a belief that the right booster may have hit the water in one piece, but may have broken up on impact. The left booster may have broken up when detonated, he surmised. Early on, the suspected booster indicated by the *Liberty Star* sonar seemed to be in one piece; it might be the crucial right-hand booster. But a submarine would have to go down with experts aboard to identify it.

On the afternoon of February 6, the Data and Design Analysis Task Force met at Kennedy Space Center headquarters and approved a request by the Coast Guard to terminate surface search activity as of 6 P.M. February 7. The recovery effort shifted to the bottom of the sea.

The Navy Moves In

Meanwhile, the *USS Preserver*, a Navy rescue and salvage ship, joined the salvage effort February 6 to commence the undersea work. With a length of 201 feet, the ship had a beam of 43 feet and draft of 13 feet. It was capable of 14.8 knots top speed and could lift 8 tons forward and 10 tons aft. It had an air diving support system and supplied divers with handheld video and 35-mm cameras.

The *Preserver* carried booms with 8 and 10 tons' capacity and compressed air diving apparatus, and was able to tow 30 tons. The big ship could be moored at two points to hover over or near objects located by sidescan sonar. From it, divers could go down to investigate. The *Preserver* was the first Navy salvage ship to arrive on the scene and supported the bulk of the hard diving. It stayed on the job for 60 days.

With the arrival of the *Preserver* came the deputy director of the Defense Department's Contingency Support Office, Dr. Dale Uhler. He set about analyzing radar, optical, and sonar data to determine what equipment would be needed for the deep salvage operation.

The Navy salvage ship, *USS Preserver*. (Navy photo.)

The Navy salvage ship, *USS Preserver*, searches on February 15–16 for sunken wreckage 16 to 18 miles off Cape Canaveral in 120 feet of water. (Peter D. Sundberg, Atlantic Fleet Audiovisual Command, U.S. Navy)

Navy commercial salvage contractors arrived at the Kennedy Space Center to confer with Navy and NASA officials on search and recovery tactics.

By February 7, the infrastructure of the undersea recovery program had been laid out. Not since the clearing of the harbors of Western Europe after World War II had the Navy undersea salvage organization faced so formidable a challenge. Heading this effort were Uhler and Captain Charles Bartholomew, supervisor of salvage, for the Navy and Colonel O'Connor for NASA. The coordination among the military services, the Coast Guard, and NASA has been described by those who worked in the salvage operation as highly effective. O'Connor determined the area to be searched and the order of recoveries while Bartholomew commanded the detailed operation of the ships.

When the search for floating debris was ended on February 7, the records of this operation showed that 150,000 square nautical miles of ocean had been searched and 12 tons of debris had been recovered. However, the underwater area search was a fraction of the surface area, about 480 square nautical miles—largely because radar and sonar had pinpointed the location of significant debris.

The Navy operated from a temporary base, a collection of vans containing office and communications equipment. "We set up our own little camp, so we were kind of independent," said Captain Bartholomew.[1] The ships were directed from the command post, one of the vans, by radio.

Mainly from force of habit, the recovery was carried out with military-style security, but the only discernible "enemy" was the press. Captain Bartholomew said: "We had to use a lot of code sometimes because the news media was listening. A lot of times we'd hear stuff on the radio before we knew about it ourselves because they were eavesdropping and reported it."

What the news media reporters were attempting to

do was to report the salvage operation with more depth and coherence than Navy and NASA handouts provided. If anything aboard *Challenger* was classified for military or national security purposes, it never was so designated.

A computer in one of the Navy vans collated information from the ships on sonar contacts and locations. "When the dust settled," Captain Bartholomew said, "we had about 700 discrete contacts, any one of which could have been the piece we were looking for. And everyone of which had to be looked at by either a diver, a submarine, or a remotely controlled vehicle. The contact tells you there's something but you don't know what it is. Seven hundred contacts is a lot, considering that you might get a half-dozen a day if you're lucky.

"The computer was invaluable in managing all these data, tracking it and helping us make cogent decisions. If I didn't have it, I probably would have had to hire about about 20 guys with green eyeshades, armbands, and sharp pencils to keep track of things for us."

Most of the underwater search area was influenced by the Gulf Stream, the Navy supervisor of salvage said. Probably the single most difficult part was dealing with Gulf Stream currents.

"You can't use divers in 3 or 4 knots of current," he said. "They just look like laundry floating in the wind. You can't use most [unmanned] submersibles with umbilicals [lines connecting these vessels to the mother ship's control unit]—normally what I use in the course of my business. They don't have the power to overcome the drag. They just get blown away, too. This is the reason we relied very heavily on submarines, which didn't have the umbilicals."

The Salvage Fleet

The fleet of undersea salvage vessels consisted of three NASA booster recovery ships, one Air Force range

---

[1] Luncheon talk with slides to the Canaveral Press Club, May 29, 1986.

The first piece of the left booster retrieved from sea bottom, March 8.

boat, five Navy vessels including the submarine NR-1, two manned submarines and their tenders from the Harbor Branch Foundation, and five salvage ships operated by contractors for the Navy.

Two of the NASA ships were the *Freedom Star* and *Liberty Star,* as mentioned earlier, both developed to retrieve boosters from the Atlantic. Each was 176 feet long with a 37-foot beam, a draft of 12 feet, and top speed of 17 knots. Both towed side-scanning sonars. The third NASA vessel was the *Independence Star,* built to recover boosters from the Pacific when shuttles begin flying from Vandenberg Air Force Base, California. It was larger than the others, with a length of 199 feet, a beam of 40 feet, and draft of 15 feet. It carried a remotely

controlled submersible called *Deep Drone,* with high-resolution sonar, three video cameras, a pair of still photo cameras, and two manipulators. *Deep Drone* could operate at a depth of 6,000 feet.

The Air Force range boat, LCU (Landing Craft Utility), had a length of 126 feet and a beam of 24 feet, with a 6-foot draft. It carried scanning sonar and was equipped to support divers at a maximum depth of 200 feet. It had a lifting capacity of 10 tons.

Like the *Preserver,* the Navy's salvage vessels were large, equipped with sonar and divers and decompression chambers. The *USS Opportune,* 214 feet long with a 44-foot beam and 16-foot draft, carried 22 divers and an air diving system capable of supporting divers to 190

The booster recovery ship, *Independence*, later the *Independence Star*. (NASA)

The Air Force Landing Craft Utility (LCU). (NASA)

The nuclear-powered Navy submarine NR-1 (NASA)

The Navy submarine rescue vessel *USS Sunbird*. (NASA)

The Harbor Branch Foundation research vessel, *Seward Johnson*. (NASA)

feet—as did the *Preserver*. The *Opportune* had lifting capacity of 20 tons forward and 12 tons aft. It carried the remotely controlled submersible *Scorpio*.

One of the big Navy salvage ships was the *USS Sunbird*, 252 feet long, with a 44-foot beam and draft of 16 feet. It had 3,440 square feet of clear deck space. It carried scanning sonar, 24 divers, and diving support with a helium-oxygen mixture to 300 feet. The *Sunbird* was equipped for submarine rescue and had 10 tons of lifting capacity and top speed of 15 knots. Its twin in size was the *USS Kittiwake*, similarly equipped with sonar and support for 24 divers. It, too, had submarine rescue capabiltiy. As was mentioned earlier, the Navy divers went down with handheld video and 35-mm cameras.

The Navy's research submarine, NR-1, was designed for wide-area search. It was 137 feet long with a 16-foot beam and draft of 15 feet. With its manipulator, it could lift 1,000 pounds and could recover a 500-pound object. It carried side-looking sonar capable of scanning a search width of 600 feet with 1-foot resolution and 2,400 feet with 4-foot resolution. It carried also forward-looking sonar with a range of 3 to 1,500 yards with 1 to 30 yards' resolution; and it had 11 video cameras and 4 still photo cameras.

The ships leased to the Navy under contract were the *Stena Workhorse*, 320 feet long with a 61-foot beam, carrying the remotely controlled submersible *Gemini*; the *G.W. Pierce II*, 158 feet long with a 30-foot beam, carrying 12 divers; the *Paul Langevin III*, 144 feet long with a 29-foot beam and 10-ton lifting capacity; the 70-foot *Eliminator* and the 90-foot *Pelican Princess*, both support vessels.

Leased also by the Navy were two manned submarines and their support vessels from the Harbor Branch Foundation. As mentioned earlier, NASA officials had consulted the oceanographic institution shortly after the accident about its recovery resources. They were available.

The Harbor Branch Foundation, an ocean research organization, is located 90 miles south of Cape Canaveral on the coast north of Fort Pierce. A number of its scientific, technical, and administrative people witnessed the *Challenger* explosion from the roof of the institution's main building, where it was customary for many to gather to watch shuttle launches. With their brilliant plumes of yellow-orange exhaust and massive clouds of smoke and steam, shuttle launches are visible for distances up to 200 miles on a clear day and can often be seen from across the Florida peninsula at Clearwater on the Gulf coast.

The foundation was organized in 1971 as a not-

The 13-ton Harbor Branch Foundation submersible, *Johnson-Sea-Link II*. Its "mother ship" is the *Seward Johnson*. (NASA)

for-profit center for marine and oceanographic research by J. Seward Johnson, Sr., heir to the founder of the Johnson and Johnson Pharmaceutical Company, and Edwin A. Link, inventor of the Link Trainer for airplane pilots, both devotees of ocean science and technology.

Harbor Branch operated a small fleet of surface and submersible research vessels with sophisticated scientific instruments, sonar, and video. The institution has two four-man submarines, three support ships capable of global reach, and a remotely operated undersea vehicle called CORD (Cable Observation Rescue Device). The support ships carry the submarines to the site of an investigation, launch them from an A-frame structure on the aft deck, and pick them up after a mission.

The two submarines, *Johnson-Sea-Link I* and *Johnson-Sea-Link II,* are 22 feet long and 10 feet, 7 inches high with a 7-foot, 11-inch beam. Each weighs 23,000 pounds. Each is battery powered with a top speed of 1.75 knots and cruising speed of 1 knot. Each carries a manipulator arm that can reach out 8 feet and lift 150 pounds. The arm enables the crew inside to attach a screw pin shackle to a piece of debris so that it can be raised to the surface by a support ship cable.

Each of the submarines has three externally mounted camera systems, including fore and aft 35-mm cameras that can take up to 800 frames per load, a 70-mm camera with a 40-mm lens, and four 400-watt strobe lamps and a high-resolution color video camera on a pan and tilt table, controlled from inside the pilot's sphere. With an attached video recorder and color monitor, the crew can see and record what the camera sees. This video camera was to prove invaluable in identifying critical pieces of the right booster.

One submarine support ship is the 176-foot *Seward Johnson* with a beam of 36 feet and 12-foot draft. It has an 18-ton sub-handling system and can lift 5 tons. It is capable of precise station-keeping with bow and stern thrusters. The other is the *Edwin Link,* 123 feet long with a 27-foot beam and a draft of 11.5 feet. In addition to 5-ton lift and a 12-ton sub-handling system, it carries a decompression chamber. The ship is a converted Coast Guard cutter.

The institution's third research vessel, the 100-foot *Sea Diver,* built for Edwin Link in 1958, carried the remotely operated submersible, CORD, on the after-deck. Director Roger Cook, a former Navy diver and demolitions expert, sent the CORD, 4 feet, 5 inches by 5 feet, 8 inches, down first to reconnoiter the *Challenger* debris area. CORD carried a video camera, still cameras, and sonar. Cook had been advised that booster parachutes and their shrouds were lying on the bottom. He wanted to see the layout before sending down a manned subersible that might become entangled in them. After a day's reconnaissance, he sent CORD back to Harbor Branch and began search operations on February 14 with *Sea-Link II.*

The Undersea Search

At midmorning Sunday, February 16, the submarine *Sea-Link II* encountered booster debris in 1,200 feet of water about 40 miles offshore from Cape Canaveral. NASA public information officer George Diller issued this memorandum:

The four-man submarine dispatched yesterday has found what is believed to be components of the right hand solid rocket booster from the Space Shuttle Challenger. Underwater photography and other data will be compared with closeout pictures taken of the 51-L space shuttle vehicle before launch to verify that this is indeed the right hand solid rocket booster. This verification will be complete after review by the appropriate engineers and technical staff this evening and Monday morning

This was the first breakthrough of the undersea search. Colonel O'Connor recalled: "We had the tracks of two solid rocket boosters, but we couldn't tell from that which was which. We had two solid rocket boosters on the ocean somewhere. We had to prioritize both boosters until we got something with a parts number that said, 'this is the right one.'"

Sea-Link II and its crew continued the videotaping

The salvage ship *Stena Workhorse* unloads a big fragment of a booster at the Trident Basin, Cape Canaveral, March 30. (NASA)

of what appeared to be parts of the right booster on February 16, making two deep dives. Both *Link I* and *Link II* carried modified Sony broadcast-quality video cameras. Director Cook said they cost about $100,000 apiece. "There are only two cameras like these in the world," he said, "and we have both of them."

On February 19, Marshall Space Flight Center engineers positively identified the debris photographed by the *Sea-Link II* crew as part of the right-hand booster. A serial number was visible on one of the pieces.

The pace of the search began to pick up. NASA's *Independence Star,* operating in 210 feet of water, launched its remotely controlled submersible *Deep Drone*. The sub's video camera detected debris that looked like a part of a booster. *Sea-Link II* shortly thereafter filmed some wreckage believed to be from the left booster. The left booster had a relatively low priority, however.

On February 22, *Sea-Link II* located a piece of the external tank, 15 by 24 feet, and attached a shackle to it. It was lying in 100 feet of water 25 miles off the coast. The *Seward Johnson* hauled it on deck and brought

it to Port Canaveral. *Deep Drone* then photographed a part of an orbiter main engine, and the *Independence Star* picked it up from 90 feet of water. It had been lying in the western part of the search area, about 20 miles offshore.

Working in the northeastern sector of the search area, the Navy's NR-1 photographed a piece of debris 12 feet long that looked like part of a booster. It was 35 miles offshore in 800 feet of water.

In early March, high winds and seas reduced the salvage effort to nearly zero on some days, but Navy divers operating from the *Sunbird* recovered two pieces from the external tank on March 6, and divers from the Air Force's LCU picked up debris from *Challenger's* left wing and aft fuselage. A day later, the LCU returned to port with a section of the forward orbiter fuselage. Divers from the vessel reported that they had found *Challenger's* crew compartment debris 15 to 16 nautical miles offshore in 100 feet of water.

The *Preserver* moved into the area, and its divers positively identified the crew compartment on March 8. Parts of it were recovered. There followed a long drawn-

Much of the offloading of *Challenger* debris recovered from the ocean took place at night after daylight searching. Here the Coast Guard Cutter *Dallas* offloads sections of the orbiter at the Trident Basin. (NASA)

out and secretive search for the bodies of the crew. Details of this process were not communicated to the news media, nor were they made public later by the Presidential Commission.

Meanwhile, the *Stena Workhorse* recovered a 7-by 6-foot section of the aft segment of the left booster. The hunt went on for the aft segment of the right booster, which was the crucial one.

On March 12, the *Preserver* returned to port with crew compartment debris. Among it were the orbiter's five general-purpose computers that fly the ship. Later analysis showed that the computers were beginning to gimbal the main engines to compensate for the loss of thrust in the right booster just before the fireball erupted. Three flight recorders also were recovered from the compartment wreckage.

Following a bout of stormy weather March 13, salvage operations were resumed on the 15th and 16th. The *Stena Workhorse* recovered a booster section weighing 3,250 pounds. It was later identified as the aft

portion of the forward segment of the right booster. A smaller section, 5 by 6 feet, weighing 560 pounds, was recovered on the 17th. It was the right booster aft segment.

The salvage effort was approaching its priority goal—the hardware that would confirm the film scenario of a leak in the joint between the aft and aft-center segments. On March 19, the search area was extended. Main engine debris was found. The *Stena Workhorse* picked up another section of the right booster 10 by 14 feet, weighing 3,000 pounds. Then *Sea-Link II* located five small pieces, one of which was the nose cone with two orange parachutes.

On March 20, a gyroscope assembly with attached cable was recovered. A part number identified it as part of the right booster forward skirt. It was found in a debris field 42 nautical miles offshore by the *Stena Workhorse*.

Colonel O'Connor appeared before the Presidential Commission in the State Department Auditorium on

The submersible *Scorpio* worked with its mother ship, the *USS Opportune*. (NASA)

March 21 and presented a salvage progress report. Despite intermittent bad weather in March, the salvage team expected to bring up more components of the right booster in the next few weeks. He said that most of the right booster components and some of those of the left booster were being found fairly close to the axis of the Gulf Stream in 220 and 1,200 feet of water.

The recovery, he said, was being worked with 9 ships, 1 manned submersible (the *Sea-Link II*) and 27 divers. A second Harbor Branch submarine, the *Sea-*

Underwater photo of a hydraulic reservoir from the aft skirt of one of the solid rocket boosters. It contained about two gallons of hydraulic fluid serving the thrust vector control (power steering) system. (NASA)

Link I, was being added to the fleet, replacing the Navy's NR-1, which had been released temporarily. O'Connor said the Navy would bring it back for deep-water searching if necessary. He described the *Sea-Link II,* whose crew had first located the right booster pieces, as "one of our most successful vessels."

O'Connor reported also: shuttle components were found at 29 locations; 20 percent of the orbiter had been recovered; parts of the orbiter's main engines had been brought into port; about 65 percent of the inertial upper stage (IUS) rocket but only 1 percent of the Tracking and Data Relay Satellite it was to boost to geostationary orbit had been recovered; 10 percent of the external tank and the same percentage of both solid rocket boosters had been recovered. Although the main priority elements of the right booster were still on the ocean floor, the salvage team knew where they were.

"We have formed a recovery team specifically to support the recovery of the right booster components," O'Connor said. "This team is comprised of NASA engineers from both Kennedy Space Center and Marshall,

a contractor team from . . . Thiokol and some of their other support contractors. We have the design team in place to support all of our activities as quickly as possible when we recover the components. The identification of the critical hardware is being provided by this team."

Chairman Rogers asked O'Connor if he was able to estimate how long it would take to complete his work. Assuming good weather, O'Connor said, most of the significant components of the right booster would be raised in two to three weeks. Some portions of the center segment of the booster had been identified, and beyond them lay a debris field that was believed to include most of the aft segment and aft skirt. O'Connor said that recovery of aft segment debris would begin the following week.

The aft segment had broken into many pieces, presumably when the booster was detonated by the range safety officer. The pieces were scattered, and it would take some time to pick them all up. The *Stena Workhorse* would be used to recover these fragments.

It was an encouraging report. Still, Rogers was exhibiting some concern about getting the evidence in time to allow him to complete his report to the President. Conclusive data had to be in hand by the end of April to enable him to do this, and March was nearly gone.

Stormy weather interrupted salvage operations starting on March 22, and although the *Stena Workhorse* remained at the right booster recovery site, other vessels remained in port. On March 28, the *Stena Workhorse* stopped hauling up debris when its submersible *(Gemini)* lost its tether and fell back into deep water. The remotely controlled submarine had continued to search under water, but turbulence had made visibility near zero.

The search fleet resumed recovery operations March 29. Recalled to help, the NR-1 cleared 13 sonar contacts, one of them identified as shuttle debris. During this period, the *Preserver* continued picking up fragments at the crew compartment site.

A Most Reasonable Approach

During the first week of April, the salvage team hired a scallop boat, *Big Foot,* to dredge the site of the crew compartment in an effort to speed up recovery of the bodies of the crew.

Over a period of days, the remains of crew members were brought up by the *Preserver,* having been placed in containers by the ship's divers. They were offloaded at night secretly and taken by ambulance to the Life Sciences Support Facility in Hangar L on Cape Canaveral, where preliminary forensic analysis and identification were done. The remains were then removed by hearses to the Shuttle Landing Facility, where they were placed aboard an Air Force C-141, They were flown to the Air Force Military Mortuary at Dover, Delaware for final processing before being released to their families.

Colonel O'Connor explained the policy of secrecy in these terms: "It was decided that the most reasonable approach to that was to say nothing until the [recovery] process was complete. There were several areas of concern in that respect. One—that if we made it visible in any sense at all, then there would be the desire on the part of a lot of people to go out there and observe the operation. So we said—for the safety of our divers and our other assets out there we'll just close the area.

"NASA astronauts were out there when we first discovered we had the crew compartment. Many engineers were involved in it. So the necessary information to assess crew safety was captured in real time. They weren't excluded from the process. Excluded were the press, people not involved in recovery activity, nonessential personnel. The families were consulted on this through the astronauts who were assigned to them in a liaison function, and rather than have a big deal about it, it was just a lot cleaner and a lot better for the families, I think for the nation, to make that ultimate one release very close and right to the world and then follow it up with the proper ceremonials."

Sonar Contact 131

The *Sea-Link I,* supported by the research vessel *Edwin Link,* scanned the crew compartment site with video tape and assisted in the recovery of small debris until April 7. Coast Guard divers working off the *G. W. Pierce* went down to investigate sonar contacts at the crew compartment site. Videotaping of the crew compartment debris was continued on April 12 by the remotely controlled submersible *Deep Drone,* working from the *Independence Star.*

Earlier in April, the NR-1 investigated a sonar contact identified as number 131. Captain Bartholomew related: "It was in an area way to the south of where we thought the right booster parts were, so we had had a very low priority on it. We were doing the whole area so when they [the NR-1] came through there, they picked this up [on video tape]. The Morton Thiokol representative didn't recognize it as being a booster

The airlift of crew remains departed at 9:30 A.M. from the Kennedy Space Center to Dover Air Force Base, Delaware, April 29, 1986.

piece because of the angle that the submarine had. When I gave the video tape to NASA, they looked at it and said they could see something funny on this—might be a burn hole. To my simple mind there was no doubt that had been burned through by a failure of a joint. That's about a 28 by 15 inch semicircular hole. I was ready to go home."[2]

The piece of debris known as sonar contact 131 was part of the right booster's aft-center segment containing the tang portion of the suspect joint burn-through area. The first contact had been made as early as

[2] Canaveral Press Club luncheon talk, May 29, 1986.

March 1 by the *Liberty Star's* sidescan sonar in 560 feet of water. The NR-1 inspected the site on April 5 and classified it as possible *Challenger* debris. The video tape made by the Navy submarine of the piece was suggestive. *Sea-Link I* was then assigned to investigate the contact and obtain higher-quality video.

The Harbor Branch submarine made a dive on April 12. On board observers verified that the piece of debris was from the right booster and included the tang portion of the joint. The *Stena Workhorse* was then instructed to pick it up. Using the submersible *Gemini* to attach a lift line, *Stena Workhorse* pulled up the piece on April 13. Its mass was estimated at 2,400 pounds.

One piece of the "smoking gun" debris that proved the theory of a solid rocket motor burn-through as the cause of the *Challenger* disaster is shown here after it was recovered from the bottom of the Atlantic Ocean. It was designated as piece number 712 and reveals a portion of the burned-out area in the right-hand booster in the lower right corner. (NASA)

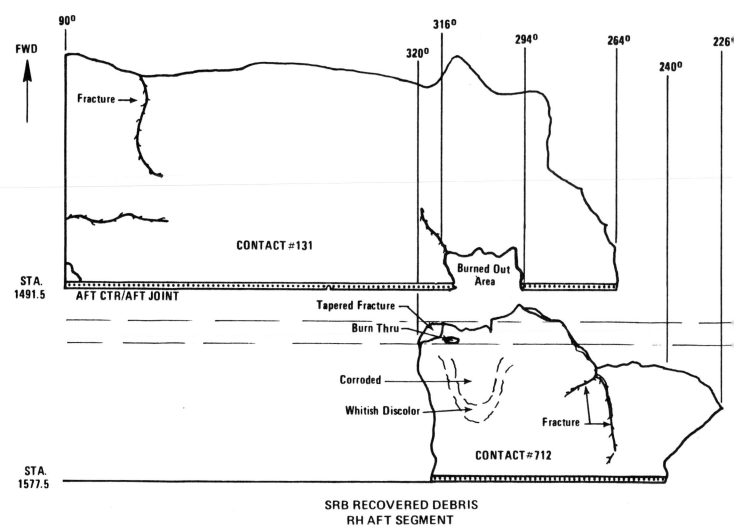

SRB RECOVERED DEBRIS
RH AFT SEGMENT
View From Outboard

This sketch in the Data and Design Analysis Task Force report submitted to the Rogers commission shows the relationship of piece 712 with another part of the burn-through area, piece 131.

The piece had a circumference of 19.94 feet and varied irregularly in height from about 6 feet to about 11 feet (71 to 132 inches). The base contained the tang portion of the lower field joint, and unburned propellant was still attached. Missing from the piece was an irregular oblong-shaped area 15 inches high and 28 inches long. Soot blistered paint around the missing area indicated that it had been burned away at a circumferential location opposite and about two feet away from the external propellant tank.[3]

[3] This was the worst possible location for a burn-through. Had it occurred 180 degrees away from the external tank, ignition of the liquid hydrogen and liquid oxygen in the tank probably would not have taken place.

On April 14, the task force notified the commission that one of two right booster sections of critical interest had been found. The other one was an adjoining piece of the aft segment that would show the lower part of the burn-through area. It was found later when the *Stena Workhorse* recovered a piece known as sonar contact 712. It was determined by NASA engineers to be a fragment of the aft segment 5 by 6 feet in extent and showed a burned-out area in the same circumferential location as that of the aft-center piece.

Examination of it at Hangar O on Cape Canaveral showed a small hole 4 by 1.5 inches that had been burned through it—not from inside to outside but from the outside in. The hole was just above a portion of the

Close-up view of piece 131, the other part of the "smoking gun" evidence of burn-through. The Data and Design Analysis Task Force called attention to the tapered edges of the casing, indicating burn-through from the inside to the outside. (NASA)

Solid rocket motor piece no. 131 from the right booster, one of main "smoking gun" pieces showing that the motor case was burned through by a hot gas leak. Note the burned-out area at the base. (NASA)

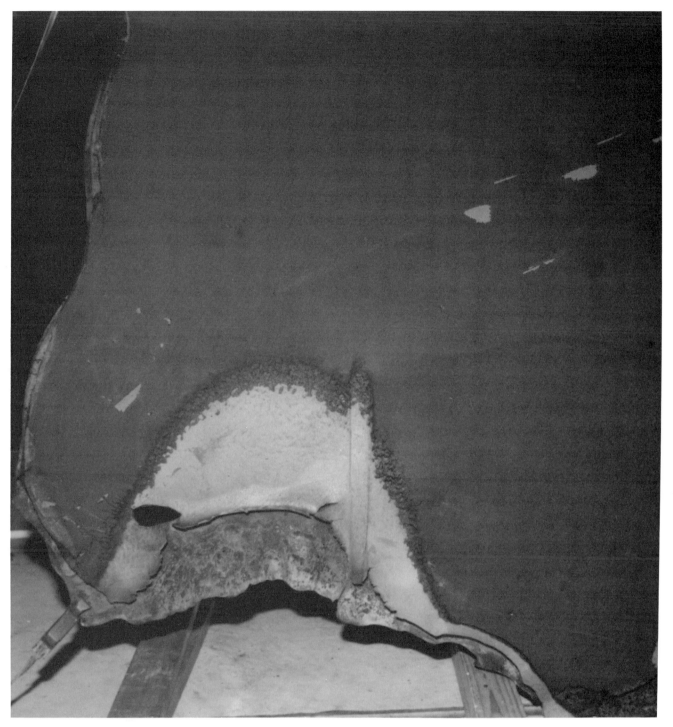

Another part of the "smoking gun," piece no. 712 from the right booster, shows a small hole *(lower left)*. View is from the inside, and the inward taper of the hole indicates that it was burned through from outside to inside. It was speculated that hot gas emerging from motor case was deflected back into the case by the external tank attachment strut. (NASA)

lower stub of the external tank attachment ring. Investigators from the Marshall and Johnson space centers and from the National Transportation Safety Board concluded that the small hole was burned by exhaust gas deflected from the main hole in the joint by the tank attachment strut.

From the dimensions of the total area missing from the two pieces, the investigators determined that 862 square inches or about 6 square feet of the joint between the aft-center and aft segments had been burned away, including 28 circumferential inches of the lower field joint.

The task force concluded that the most probable cause of the 6-foot-square hole through the motor case was a combustion pressure leak past the O-ring seals in the field joint. However, because of thermal destruction of the joint, including the seals, other possible sources of combustion pressure leak could not be excluded by the physical evidence.

The small hole burned through from the outside provide a clue to the next event in the process that destroyed *Challenger* and the crew. The aft attachment strut connecting the right booster to the tank separated, burned through by the torch of 5,600°F gas it deflected back toward the joint. Free of the lower strut, the booster yawed about its forward attachment to the tank, and the nose structure with the heavy frustrum crashed into the tank. (Thus was confirmed the scenario of the cause of the disaster as theorized from film and telemetry data months earlier.) The impact ruptured the intertank structure, allowing thousands of gallons of liquid hydrogen and liquid oxygen to ignite and erupt into a massive fireball around the shuttle.

The task force reported to the commission: "Therefore it is concluded that the partial separation of the right booster at its aft lower attachment connection to the tank and the breach of the liquid hydrogen tank led directly to the destruction of STS 51-L."

The salvage operation reached a climax with the recovery and positive identification of the burned booster fragments. Sonar contacts 131 and 712 contained 90 percent of the burn-through area. For practical purposes,

the evidence was in: the joint had failed. What was not known was how or why the seals failed; that effect could only be surmised. Recovery operations continued on a declining scale of activity.

During the latter part of April, additional pieces of the right booster were recovered, one weighing 3,200 pounds. More evidence: on April 28, the *Stena Workhorse* recovered a large part of the aft segments of the right booster, with a burned area 33 inches long. It was regarded as the final piece of the jigsaw-puzzle assembly of the failed joint. The search in deep water was logged as completed on April 30, but the salvage team continued to pull up debris that had been located by sonar and identified as important.

Titan Fails

Another blow to the American space effort came on April 18 when a Titan 34-D launch vehicle exploded 6 seconds after it was launched with a military reconnaissance satellite from Vandenberg Air Force Base, California. With the shuttle grounded indefinitely, loss of the Titan, the next most powerful launcher in the American inventory, temporarily stalled the Department of Defense satellite program pending analysis of the Titan failure.

Like the shuttle, Titan was powered by liquid and solid fuel boosters. The vehicle was manufactured by Martin Marietta; the liquid propellant engines by Aerojet; and the solid motors by United Technologies. A Titan 34-D had failed before, in August 1985, also carrying a reconnaissance satellite on a launch from Vandenberg. That failure had been diagnosed as premature shutdown of the liquid fuel engines in the first stage. The 1986 explosion was attributed to a solid rocket booster failure. Thus, both propulsion systems on the Titan had failed catastrophically in less than a year.

Like the shuttle boosters, the Titan solid rocket motors were stacked in segments, but sealed with only a single O-ring instead of the two used in the shuttle SRB. In testimony before the Rogers commission, Mar-

At 6:18 P.M., May 3, 1986, Delta launch vehicle 178 rose from Cape Canaveral with the Geostationary Operational Environmental Satellite (GOES G) aboard for the 1986 hurricane watch. (NASA)

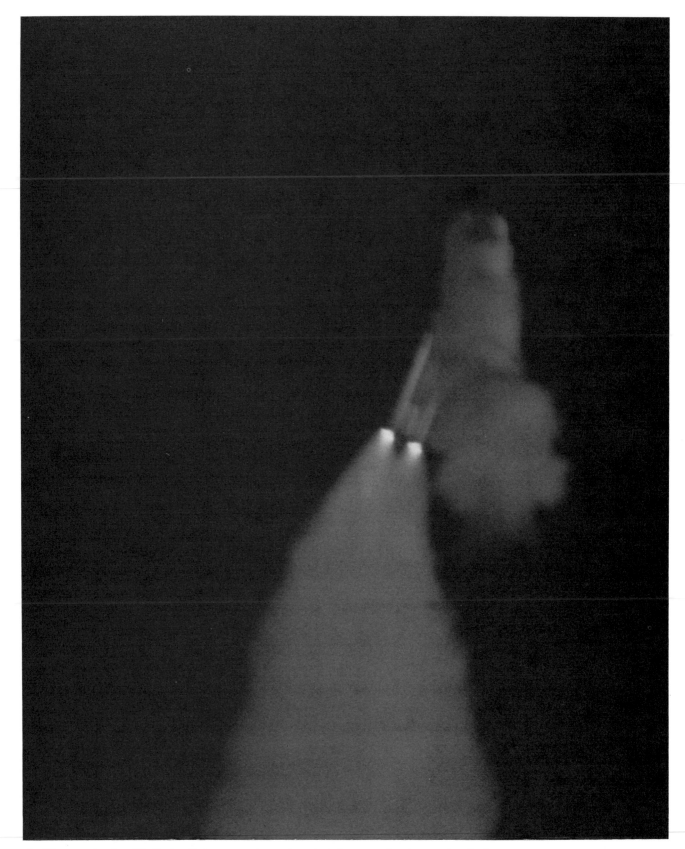

At 71 seconds after liftoff, the first-stage main engine shut down, owing to an electrical failure, and the Delta began to tumble. The range safety officer sent a signal to destroy it. (NASA)

Once more, debris filled the skies over the coast of Cape Canaveral. Another GOES weather observer was launched successfully by a Delta from Cape Canaveral, February 25, 1987. (NASA)

shall engineers had cited the consistently good performance of the Titan boosters with only one O-ring as a convincing indication that the shuttle boosters were safe with two, even if the secondary seal was characterized officially as nonredundant. Now that argument had gone up in smoke.

Delta

The United States space program suffered its third blow on May 3 when a NASA Delta 3920 rocket carrying an 1,851-pound weather satellite failed 70 seconds after launch from Cape Canaveral and was destroyed by the range safety officer. It was the 178th launch of the Delta, NASA's principal expendable launcher for the last quarter of a century.

The Delta payload was GOES 7 (Geostationary Operational Environmental Satellite). It was targeted for an equatorial orbit at 22,300 miles' altitude. There it would hover relatively stationary at 75° west longitude, where it could photograph weather systems and track hurricanes in the Atlantic Ocean and the Caribbean Sea.

Not only was the loss of GOES 7 a serious handicap for the National Oceanic and Atmospheric Administration and its parental Department of Commerce, but the destruction was the first major Delta failure in years and confirmed fears that quality control had gone awry in American rocketry. Now the Delta was grounded like the shuttle and the Titan until the causes of these failures could be determined and fixed. Only one major satellite carrier system remained operational in the United States: the Atlas Centaur.

Foreign satellite launch services were reported to be moving in to fill the gap, although Ariane, the French-launched European Space Agency vehicle, was also having its troubles.[4] Now China and the Soviet Union announced they would enter the commercial satellite launching market, and the Japanese space agency indicated it would soon be in a position to compete.

The Delta rocket debris had barely reached the ocean when the shuttle salvage vessels, engaged in cleanup work, steamed after it. The *USS Opportune* began a search for electronic boxes, wiring, and other electrical parts from the rocket's midsection. Although its small solid propellant rockets (Morton Thiokol) had functioned nominally, the first-stage liquid fuel engine (Rocketdyne) had shut down prematurely, as if switched off. Telemetry clues suggested an electrical failure. With loss of thrust and attitude control, the 115-foot rocket began to tumble, and range safety signaled the destruct package to explode.

Recovering the Delta debris was a relatively straightforward task compared with the *Challenger* salvage. Colonel O'Connor said: "When the Delta went in, within three days of operations, we had up what they wanted. Because we had everything in place and knew how to do it." The *Challenger* salvage data, including photographs, sketches, underwater pictures, schedules, and text, made up a package weighing 400 pounds. The package described the recovery of 86,000 pounds of shuttle wreckage including 42 pieces of booster debris. "Delta showed why it was really important to capture all this," said O'Connor.[5]

The Orbiter

The task force reported that 30 percent of the *Challenger* structure was recovered during the three months' salvage effort. Many pieces shown by underwater video were not recovered because they were not deemed important in establishing the cause of the ac-

---

[4] On May 30, Ariane 2 was blown up by a range safety officer minutes after it was launched from Kourou, French Guiana. The third stage failed. It was carrying an Intelsat V F-14 communications satellite.

[5] The rapid recovery of the Delta wreckage facilitated the identification of an electrical short circuit as the cause of the Delta failure. A second Delta (3920) was successfully launched from Cape Canaveral September 5, 1986 with a "Star Wars" experiment that resulted in the destruction of a simulated Soviet reentry vehicle.

With shuttle salvage just completed, resources were available at Cape Canaveral to recover the Delta wreckage promptly. Here the first-stage main engine of the rocket is unloaded from the Navy ship *Opportune* after being picked up 30 miles east of the Cape in 160 feet of water. (NASA)

cident. The search, however, took hundreds of thousands of hours of work.

Significant pieces of the orbiter that were recovered included all three main engines, the forward fuselage including the crew module, the right inboard and outboard elevons, a large part of the right wing, a lower portion of the vertical stabilizer, three rudder-speed brake panels, and portions of the midfuselage side walls on left and right sides.

All the fractures and failures of the structure, except for the engines, were concluded to have been caused by overload forces. There was no evidence of internal

burn damage or exposure to explosive forces, the task force reported. The destruction of the orbiter was the result predominantly of aerodynamic acceleration and inertial forces that exceeded design limits. During the breakup, the right booster struck the outboard end of the orbiter's right elevon. The right side of the orbiter was probably sprayed by hot propellant gases exhausting from the hole in the right booster joint, the report added.

As mentioned earlier, the crew module broke away from the fuselage and emerged intact from the fireball. Although its fall is not specifically described in the

Here the boat tail or bottom of the Delta main stage, recovered with the main engine, is unloaded from the *Opportune*. (NASA)

commission report, the radar data the commission received indicates that when the fireball erupted, at about 74 seconds, it was at 50,800 feet altitude, and at the time of the breakup of the whole vehicle, the altitude was 104,000 feet. Apogee of the breakup debris was 122,400 feet. Whether the crew module reached that altitude was not indicated in the report.

The task force and commission did not make public the pathway described by the crew module after it was detected flying out of the fireball, but the question arises of whether such a module equipped with a parachute descent system could save a crew of equal number in similar circumstances. As mentioned earlier, an excla-

mation heard on the recovered voice recorder has indicated that the pilot, Mike Smith, was aware something had gone wrong. It can be assumed that the crew survived the fireball, but whether they perished by decompression or impact was not established. The crew module was found submerged in about 90 feet of water in an area of 20 by 80 feet. There was no evidence of an internal explosion, the report said, nor of heat or fire damage on the crew module fragments or on those of the forward fuselage. The module was disintegrated, with the heaviest fragmentation and crush damage on the left side.

The report said that the fractures examined were

typical of overload breaks and appeared to be the result of high forces generated by impact with the surface of the water.

The report went on to say that consistency of damage to the left side of the outer fuselage shell and the crew module indicates that these structures remained attached to each other until impact with the water. Evidence that the crew module remained "an essentially intact structure until impact with the water was the concentration of pieces in only one small area." The module is supported within the forward fuselage at four primary attachment points and is welded to create a pressure-tight vessel.

Approximately 75 percent of the module was recovered. The internal components of the module were crushed and distorted. The magnitude and directions of the crush damage indicate, the report said, that the module was in a nose-down and steep left-bank attitude when it struck the water.

Dr. Kerwin's Report

The investigation into the cause of death of the crew was headed by Dr. Joseph P. Kerwin, director of Life Sciences at the Johnson Space Center. His report to the commission suggested that the crew survived the fireball and the breakup of the orbiter and perished of either decompression shortly thereafter or impact with the water.

Dr. Kerwin reported that forces at breakup of the orbiter were probably too low to cause death or serious injury. The forward fuselage was expelled from the disintegrating external tank and pitched nose down. He estimated there was a brief period of acceleration ranging from 12 to 20 g's, then a rapid drop-off to 4 g's in less than 10 seconds. He stated that these accelerations are survivable, and the probability of injury is low.

When the crew module broke away from the fuselage, the crew's oxygen supply from the orbiter normally provided only when they wore helmets was cut off. However, crew members could continue breathing air—not pure oxygen—from emergency personal egress air packs connected to their helmets. The pack had two cylinders each containing 0.73 liters (about 3/4 of a quart) of ambient (cabin) air at 2,100 pounds of pressure per square inch. The supply would last about six minutes to allow the crew to make a quick exit from the cabin in case of an emergency on the pad.

The individual air packs had to be turned on manually. Four air packs were recovered. Three had been activated. One, identified as Mike Smith's, was found to be three-fourths to seven-eighths depleted. Another air pack, similarly depleted, could not be identified. Nor could a third, which appeared to have been activated; the extent of its depletion was not reported. The fourth air pack was not activated and was identified as belonging to Dick Scobee, the commander.[6]

The partial use of the air packs suggests that explosive decompression in the crew module did not take place at breakup. If it had, Dr. Kerwin said, the fact that the crew were not wearing pressure suits would have caused them to lose consciousness within 6 to 15 seconds. On the other hand, if the module had not lost pressure immediately, or had lost it gradually, the crew might have remained conscious longer or until ocean impact.

Dr. Kerwin said that he and his colleagues sought to learn whether explosive or gradual decompression had occurred in the crew module before impact: "Much of our effort was expended attempting to determine whether a loss of cabin pressure occurred. We examined the wreckage carefully, including the crew module attach points to the fuselage, the crew seats, the pressure shell and flight deck and middeck floors, and feedthroughs for electrical and plumbing connections.

---

[6] The air pack activation switches for Scobee and Smith were on the backs of their seats, and they would have to climb out of their seats to turn them on. In its report of August 4, 1986, *Aviation Week and Space Technology* interviewed astronauts who explained that either Judith Resnik or Ellison Onizuka turned on Smith's pack. The air packs of other crew members were located beside their seats.

A 70-mm tracking camera south of pad 39B picked up this cone-shaped object *(center)* emerging from smoke plumes at 11:39:29 Eastern Standard Time, January, 28, 1986. It was later identified as the crew compartment of the *Challenger* orbiter. (NASA).

The object is identified as the orbiter crew cabin in this NASA photo published in the *Commission Report,* vol. 3, p. N-77.

"The windows were examined and fragments of glass analyzed chemically and microscopically. Some items of equipment stowed in lockers showed damage that might have occurred due to decompression. We experimentally decompressed similar items without conclusive results. Impact damage to windows was so extreme that the presence or absence of in-flight breakage could not be determined. The estimated breakup forces would not, themselves, have broken the windows. A broken window due to flying debris remains a possibility."

In his report to Admiral Truly, Dr. Kerwin said the findings of this team were inconclusive. The cause of death of the *Challenger* astronauts cannot be positively determined. The crew possibly but not certainly lost consciousness in the seconds following orbiter breakup because of in-flight loss of crew module pressure.

After the orbiter broke up, Dr. Kerwin's report said, the crew compartment continued its upward trajectory, peaking at an altitude of 65,000 feet approximately 25 seconds after breakup. It then descended, striking the ocean surface about 2 minutes and 45 seconds after the breakup at a velocity of about 204 miles per hour. The forces imposed by this impact approximated 200 g's, far in excess of the structural limits of the crew compartment or crew survivability levels.

Data supporting the possibility that the crew lost consciousness due to loss of crew module pressure was listed as follows:

1. The accident happened at 48,000 feet.[7] The crew cabin was at that altitude or higher for almost a minute. Without an oxygen supply, loss of cabin pressure would have caused rapid loss of consciousness, and it would not have been regained before water impact.

2. Personal air pack activation could have been an instinctive response to unexpected loss of cabin pressure.

3. If a leak developed in the crew compartment as a result of structural damage during or after breakup, breathing air available would not have prevented rapid loss of consciousness.

4. Crew seats and restraint harnesses showed patterns of failure that demonstrate that all seats were in place and occupied at water impact, with all harnesses locked. This would likely be the case had rapid loss of consciousness occurred, but it does not constitute proof.

The report concluded: "Impact damage was so severe that no positive evidence for or against in-flight pressure loss could be found. Finally, the skilled and dedicated efforts of the team from the Armed Forces Institute of Pathology and their expert consultants could not determine whether in-flight lack of oxygen occurred nor could they determine the cause of death."

Exemption 6

What killed the *Challenger* seven? Was it catastrophic decompression when explosive forces hurled the crew compartment clear of the flaming wreckage— or impact with the ocean?

The inability of the Kerwin report, relying on au-

[7] Radar data, as mentioned previously, put the "mishap" at 50,800 feet.

topsy findings by the Armed Forces Institute of Pathology, to resolve the probable cause of death is the missing link in the *Challenger* investigation. It has a bearing on an eventual space agency decision about a launch escape system.

If some or all crew members survived the breakaway from the fireball, a possibility considered by the report, the probable cause of death was ocean impact. In that case, a crew module equipped with escape and parachute descent systems would have saved them— and averted the most horrendous result of the accident.

The question of whether such systems are feasible for the orbiter is not one of engineering, but of policy. Installing an escape module would require redesign of the orbiter. It would have serious impacts on shuttle program costs performance, and schedules.

But so long as the probable cause of death is not determined, there remains doubt that any launch escape system could preserve the crew in an accident like the one that destroyed *Challenger*. NASA has denied public access to the autopsy reports, including access by the American Medical Association. Dissemination of the autopsy results beyond government files may lead independent experts to a more definite opinion than that presented to the Rogers commission. That would be relevant to the public's interest in making the shuttle failsafe through all phases of its ascent to orbit.

Citing Exemption 6 of the 1977 Freedom of Information Act, NASA refused the request of the *Journal of the American Medical Association* to release the autopsy findings for possible publication in the *Journal of the AMA*. Exemption 6 allows the government to withhold information from personal and medical files that would constitute "a clearly unwarranted invasion of personal privacy."

NASA's refusal was appealed on September 12, 1986 by Philip E. Gunby, editor and director of the Medical News and Perspective Department of the *Journal of the AMA*, on grounds that the public's interest in the future of the space program was directly involved. In the agency's response, January 15, 1987, Ann Bradley, NASA associate deputy administrator, stated that the agency's initial determination found that a generalized public interest was insufficient "to outweigh the considerable invasion which release of the autopsy reports would inflict on the families." Further, she added: ". . . NASA's own Medical Director has publicly concluded that all scientific tests and evidence including the autopsy reports failed to establish either the time, manner or cause of the astronauts' deaths. Given this,

it seems highly unlikely that any public interest in future space activities would be served in any way by the release of the astronauts' autopsy reports."

"Therefore," she advised the editor, "none of the public interest asserted in your appeal letter tips the balance toward release."

The response concluded that this was the agency's final determination.

8. Views from the Flight Deck

Until the solid rocket booster field and nozzle joint problems were disclosed by the *Challenger* investigation, none of the men and women who flew the shuttle was aware of them and their catastrophic implications. NASA's chief astronaut, John W. Young, reacted vigorously when this unsafe condition came to light. He sent a memorandum to headquarters about it expressing his indignation and doubtless that of the entire Corps of Astronauts. It was dated March 4.

Young's memorandum landed on the desk of his Navy colleague, Rear Admiral Richard H. Truly, now NASA associate administrator for space flight. Truly did the right thing, in the view of the press. He made the memo public. It was the appropriate tactic in the context of the accident investigation, inasmuch as Young had sent copies to all the astronauts, as was his duty as chief of the astronaut office.

"From watching the Presidential Commission open session interviews on television," said the memo, "It is clear that none of the direct participants have the faintest doubt that they did anything but absolutely the correct thing in launching 51-L at every step of the way. While it is difficult to believe that any humans can have such complete and total confidence, it is even more difficult to understand a management system that allows us to fly a solid rocket booster single seal design that explosively, dynamically verifies its criticality 1 performance in its application. . . . There is only one driving reason that a potentially dangerous system would be allowed to fly: launch schedule pressure."

Young went on to describe "several other potentially dangerous examples that you can be sure were accepted for the very same reasons. Unlike the secret seal, which no one that we know knew about, everyone knows about these items." He referred to inadequate brakes and landing gear tires and the perils of landing at the Kennedy Space Center where rain and high crosswinds are frequently unpredictable.

Attached to Young's memo was another from Stephen C. Bales, chief, systems division, Johnson Space Center. It identified 34 safety-related items and 7 studies of the orbiter landing system, the external tank, the main propulsion system, the reaction control system, the fuel cell power system, the remote manipulator system, and the orbital maneuvering system.

"The enclosed memorandum by Steve Bales' division is an outstanding and incomplete list of systems, safety related items that ought to be fixed so that we do not lose any more space shuttles and flight crews. You will note that many of these potentially serious conditions were discovered after we started operating. On an individual basis, they were not big enough to slow or stop the launch rates. But totally, this list is awesome. The list proves to me that there are some very lucky people around here."

On April 3, 1986, the commission summoned the leaders of the Corps of Astronauts to a hearing in the Dean Acheson Auditorium of the State Department to hear their views on shuttle safety. To avoid any conflict of roles, Dr. Sally Ride, a shuttle astronaut as well as a commission member, declined to take part in the discussion.

It was a historic occasion. For the first time since the establishment of NASA in 1958, the astronaut leaders had the floor to tell an independent panel of experts assembled by order of the President of the United States what was amiss in NASA's stewardship of manned space flight and what should be done about it.

Here was a group of the most experienced astronauts in the world, having flown in Gemini, the Saturn 5–Apollo to the Moon, the Skylab space station, and the shuttle. Their expertise and academic credentials were on the same level as those of commissioners. But until now, while NASA managers had often appeared before congressional committees, the astronauts had never been consulted as a group about the management of the shuttle program, especially the flight rate, which some perceived as unrealistic.

In theory, the astronaut commanding a mission could say whether it would fly or not if a safety concern arose. In practice, this option depended on the flight commander being aware of the concern. On *Challenger* 51-L, the commander, Dick Scobee, halted the countdown when he became aware of a faulty door latch. But no one told him that the O-ring seals on which the safety of the spaceship and the lives of the crew depended might not function in cold weather. Why not? Why was the engineering concern about the seals not communicated to the crew? It was this gap in communication that the commission began to explore on April 3.

The witnesses included George W. S. Abbey, director of flight crew operations, a managerial function at the Johnson Space Center. Abbey, a graduate of the U.S. Naval Academy, had been an Air Force pilot for 13 years. He had a master's degree in electrical engineering and had joined NASA in 1964, working through the Apollo and shuttle programs.

John Young, retired Navy captain, was Abbey's counterpart as chief of the astronaut office since 1975.

A graduate of the Georgia Institute of Technology in aeronautic engineering, Young had joined the corps in 1962, had flown Gemini 3 with Virgil I. (Gus) Grissom; had flown Gemini 10; then Apollo 10 to lunar orbit; Apollo 16 to a landing in the lunar Descartes highlands; then he commanded the first flight of *Columbia* and the ninth shuttle mission, on which *Columbia* carried the first spacelab.

Young's deputy in the astronaut office was Paul J. Weitz, a Navy pilot who joined the corps in 1966 and flew the first Skylab mission in 1973. Weitz commanded the first flight of *Challenger* (STS-6) in 1983. Appearing as a witness with Abbey, Young, and Weitz was Navy Captain Robert L. Crippen, a graduate of the University of Texas in aerospace engineering, who had joined the corps in 1969. He flew as pilot with Young on *Columbia*'s first flight and later commanded the seventh (STS-7), the eleventh (41-C), and the thirteenth (41-G) shuttle missions. Crippen was assigned to command the first shuttle flight out of Vandenberg Air Force Base. He served as a deputy to George Abbey and had a leading role in the NASA task force investigating the *Challenger* accident.

The fifth witness was Henry W. Hartsfield, a fighter pilot in the Air Force who had joined the corps in 1970. He had a master's degree in engineering science from the University of Tennessee. He was the pilot of *Columbia*'s fourth mission (STS-4) and commanded the twelfth shuttle mission, 41-D, and the twenty-second, 61-A.

A Question of Push

Chairman Rogers opened the interrogation by asking John Young whether the astronaut office believed that it was being pushed too hard to meet the 1985 flight schedule. Young responded, at first, in the tactful, public-hearing response mode that skirts criticism but implies it. "It was hard for me to see how we could do a lot more with our people unless we do something different," he began.

"In other words," said Rogers, "you thought the activity in 1985 was about all you could handle . . . but that the pressure in 1985 was not too great. Is that correct?"

Young replied neutrally: "I thought 1985 was a really outstanding year for the space program."

Rogers persisted: "But that if you had to do more that year, it might have been too much?"

Young, finally: "I think we would have been pushing it. Yes, sir."

The real question was, had NASA overextended its resources of people and machines to meet an arbitrary schedule at the expense of crew safety?

Commissioner Robert B. Hotz, retired editor of *Aviation Week and Space Technology*, pressed the point: "John, in view of your statement that '85 was about as hard as you could push the system with 9 flights, how do you view the 15 launches scheduled for this year as far as the load on your system?"

"It's really hard for me to assess it from where I sit, but I think it would have been pretty tough," Young said.

In the background of this exchange was the understanding of nearly all those present that NASA's goal was 24 launches a year with four shuttles. But irrespective of the system's flight capability, the goal was constrained by the maximum rate of external tank production, which was 24 tanks a year. However, with a launch rate of two flights a month, the space agency's planners were confident the system could meet commercial, scientific, and military demands and also begin hauling matériel into orbit to build the space station.

Now, it was becoming clear that with a corps of 54 trained astronauts and four shuttles, the system would have been squeezed beyond a safe limit to attempt 15 flights a year. After flying 9 in 1985 and 2 in 1986, it collapsed.

Passing the Word

Chairman Rogers next asked Paul Weitz how information about joint problems in the boosters would be communicated to the astronauts. Weitz suggested that question should be addressed to Arnold Aldrich, the shuttle boss at the Johnson Space Center. The following dialogue ensued:

Rogers: In other words, you wouldn't know how that would get—and I'm trying to find out how that information gets to the astronaut community, and I gather in this case, it didn't. . . . and information about the joint that failed and, we think, probably caused the accident was not known to any of you gentlemen, as I understand it.

Weitz: That is true, which means, therefore, that if it surfaced, either we were not made aware of it or we did not realize the significance of the item. . . .

Rogers: Well, I gather so far in our investigation that . . . none of these groups knew about—at least you didn't know about it. That is correct, isn't it? None of you knew about the problems that you had been having with this joint?

Weitz: Yes, sir. That is correct.

Rogers: How did it happen that the astronauts, who are so vitally concerned with safety aspects, didn't know about this problem?

Weitz: That is part of what we're trying to reconstruct also.

Commissioner Hotz asked whether Weitz believed the joint problem should have surfaced in the flight readiness reviews. Weitz replied that he could not see "how we could say otherwise, since it turned out to be a fatal flaw."

Robert Crippen then pointed out that the astronaut office is represented at flight readiness reviews, but the seriousness of the seal problem was not recognized.

"One of the prime things we do in a flight readiness review is review the anomalies that have occurred on previous missions and decide how we have to resolve those or why we think it is acceptable to go fly with those," Crippen said.

He said that he represented flight crew operations at the flight readiness review for the mission that followed the 51-C *Discovery* launch (January 24, 1985).[1] The soot and blow-by found in 51-C were presented at the review, he said, adding: "In truth, from my perception it wasn't considered that much of a big deal and it wasn't like we had a major catastrophe awaiting in front of us. I guess the emphasis was not such that one would think that it was a major problem that it was."

Rogers: Who presented the anomaly?
Crippen: That was presented by Marshall Space Flight Center. In going through their stuff on the solid rocket boosters, it was presented as an anomaly. If my memory serves me right, we had had a putty change on the joint just prior to that, and it was alluded that perhaps the putty modification may have had something to do with that, but that it really wasn't that big of a deal.

Rogers observed that the flight readiness review's resolution of the erosion and blow-by on 51-C stated that the risk was acceptable "because of limited exposure and redundancy on the seal." He asked whether Crippen knew that the criticality 1 list showed there was no redundancy on the seal.

"I did not," Crippen replied. "And I was not aware of the waiver that had changed the joint from a 1-R to a 1. If I had been aware of that in association with the setting, I would have taken the problem much more seriously."

John Young commented: "I don't recall anything coming up on the flight readiness review on the solid rocket motor seals. . . . and if anybody in the gang had known about this business and understood it we might have said something, but really it should have been

taken care of by the process long before it ever got to flight readiness review."

Chairman Rogers raised the question of whether any astronaut office representative was aware of the effect of cold weather on the O-ring seals on the day before the launch of *Challenger* on mission 51-L.

"We were not aware, no, sir," said Weitz. "We were not aware of any concern at all with the O-rings, let alone the effect of weather on the O-rings."

Dr. Albert D. Wheelon, a physicist and senior vice president of Hughes Aircraft Company, asked the astronauts whether they sensed any particular urgency to launch *Challenger* on the flight day. Young said there is an urgency to proceed with every launch "once you get a vehicle loaded and on the launch pad." Weitz said that he did not perceive any greater sense of urgency to launch 51-L than with the others.

Crew Escape

Major General Donald J. Kutyna, director of Air Force Space Systems and Command Control and Communications, then addressed the question of a launch escape system. At this stage of the investigation, it had been widely assumed by the aerospace community that the *Challenger* crew had perished instantly in the fireball or in the breakup of the orbiter. But film evidence that the crew module emerged from the fireball and breakup intact opened up a new line of inquiry about the feasibility of launch escape.

"In the design stage," General Kutyna said, "the shuttle had several crew escape and survivability features that were contemplated but for one reason or another weren't put on the vehicle that we have today. In view of our experience, what crew survivability and escape provisions would you like to see on today's shuttle?"

Henry Hartsfield said that he would like to see a low-altitude escape system—some ability to bail out of

---

[1] Crippen identified the mission being reviewed as 51-E, but it was redesignated as 51-D and launched April 12, 1985.

Major General Donald J. Kutyna, director of Space Systems and Command Control and Communications for the Air Force, listens to astronauts discussing landing problems with the orbiter. (NASA)

the vehicle. The orbiter has contingency abort modes in case two (of its three) engines fail, he said, but the end result of that is ditching.

"When you talk about smacking the water at 200 knots with an airplane that is basically an airliner-type design, I'm convinced it's going to break up," he said. "If you've got a 60,000 pound payload behind you, it's probably going to come into the cockpit with you."

John Young said that on numerous occasions, "we have talked to people about doing things like putting in bail-out systems, such as the tractor rocket system[2] which is plain bail-out, and they have always seemed to be more than people could put up with. . . . if we don't do it for this one, for surely the next vehicle that we develop there should be an escape system."

A launch escape system was something the com-

mission was considering, Rogers said. He asked for recommendations.

John Young said he thought it would be touch and go to attempt to add an escape system before flying again. "I guess if you put the right people on it with the right money and the right effort, you ought to be able to do it pretty darn quickly, but I'm not sure that we have that kind of capability in NASA."

Crippen said that he and Young had worked on the escape problem "long and hard." It's more than money, he said. It's a touchy problem to solve technically with the present shuttle.

"If you were going back from square zero and you went to some kind of concept like the F-111 cabin escape system, you might—and build it from the ground up, you might be able to do something like that. . . . But again, I don't know of an escape system that would have saved the crew from the particular incident that we just went through. I don't think it's possible to build such a system," he said.

[2] The tractor rocket bail-out system Young referred to used small rockets to pull individual crew persons away from the orbiter hatch so that they could descend on their parachutes without striking a wing or part of the orbiter.

Henry W. Hartsfield, Jr., as he appeared as commander of the 41-D *Discovery* flight, August 30–September 5, 1984. (NASA)

Automated Flight

Commissioner Feynman raised another survivability issue. Suppose the crew is stricken somehow at reentry and unable to function, he said. Is it possible to contrive a computer-driven reentry to save crew and mission?

Crippen said that reentry cannot presently be done automatically, nor by ground control. Automatic flight reentry was considered in designing the shuttle, he said, because some people in the program believed that the shuttle should be flown the first time unmanned. However, he said that he and Young opposed a first automated flight because there was a better chance of success with a man aboard.

Feynman said that installing an automated reentry and landing system would be "fairly simple to do," including automatic lowering of the landing gear, which

John W. Young, as he appeared as commander of the STS-9 mission of *Columbia*, the first to carry Spacelab. (NASA)

is done manually. He thought of this as an idea for backup. But Crippen was not sure. Any time such a system is installed, he said, "you have built in another failure mode."

Manual deployment of the landing gear was "a very conscious decision of ours," Crippen added; ". . . putting the landing gear down at the wrong time on this vehicle can cost the vehicle."

Young criticized the range safety package, the installation of explosives that can be detonated by radio from the ground to destroy the shuttle in the event it goes out of control and threatens to hit a populated area. "I think the range safety package, if we have to carry one, should be one that doesn't tear up the whole piece of machinery including the crew . . . if you have to have one, it should be man-rated."[3]

[3] The range safety destruct system was set up by the Webb-McNamara agreement of 1963 giving responsibility for range safety during NASA flights to the Department of Defense.

Young related that he and others had fought the range safety package since the early 1970s but were never able to have it removed. There was an unwritten agreement, he said, that when the ejection seats were removed from *Columbia,* the range safety package would also come off, but that did not happen.

"How do you feel about it now?" asked Commissioner Hotz. "Do you still feel it should come off?"

"I was sure if the vehicle is reliable enough to go where you want it to go, I think the range safety package should come off. On the other hand, they've just got some data that says it's not as reliable as it should be!"

Chairman Rogers asked whether it was feasible to build an escape system for the replacement orbiter that, he said, Congress was to consider. Weitz said that it probably was not feasible. But Young said a tractor rocket escape system could be put in without too much trouble.

Neil Armstrong, commission vice chairman, who had flown Apollo with a tractor rocket launch escape system, entered the discussion. Apollo's tractor escape system was designed to pull the crew capsule away from the service module and the Saturn 5 rocket to an altitude of 3,000 feet if a blowup impended. The crew module would then descend on its parachute system to the ocean.

"You're talking about a 'yankee' sort of system?" Armstrong asked Young.

"Yes," agreed Young. "A yankee system that takes out the top of the cockpit and lets people jump out that way."

Weitz said that basically there were three alternatives, "and these have been looked at at various times over the last several years." One was the escape module. "With the orbiter as it presently exists . . . you cannot modify the existing orbiter to accommodate an escape module."

Other than the escape module, he said, "you have some sort of rocket-assisted personnel extraction where you use, as Neil said, the yankee system, or you have some sort of bail-out system." He added that a trade-

off study was already started on these alternatives, but the escape module was out of the question.

Crippen said the only feasible system he was aware of was some kind of bail-out that could be used at subsonic speeds. It didn't seem worth the complexity to go through with it.

Armstrong asked the witnesses whether it seemed reasonable to determine whether the orbiter is actually "ditchable." It seemed reasonable to Crippen because that was an unknown, he said. But Weitz did not concur. He did not believe the orbiter could survive ditching in the ocean.

"I think if we put the crew in a position where they're going to be asked to do a contingency abort, then they need some means to get out of the vehicle before it contacts the surface of the Earth," Weitz said. But he added that the difficulties of separating from the orbiter in a bail-out were quite serious.

If you go out the orbiter head hatch, he said, you would probably hit either the orbital maneuvering engine pods (on the rear fuselage) or the high, vertical tail. If you go out the side hatch at airspeeds much in excess of 220 or 240 knots, you would hit the wing.

The escape window was pretty narrow, commented Commissioner Eugene E. Covert, Professor of Aeronautics at the Massachusetts Institute of Technology. Weitz added that the bail-out scenario he and others had considered called for people to jump out of the orbiter with no equipment other than the helmets and flight suits they were wearing in the module plus their parachutes.

A Safety Panel

Commissioner Feynman shifted the discussion to the advisability of falling back on expendable rockets to launch commercial satellites instead of lifting them to low orbit in the shuttle. In the beginning, NASA had sought to justify the cost of developing the shuttle as an economical carrier for satellites in lieu of expendable rockets. Since the *Challenger* accident, that rationale for the shuttle was fading away. The shuttle had never achieved cost-effectiveness as a satellite carrier as its developers had predicted in the 1970s. Now with the fleet reduced to three vehicles, the orbiter was considered too valuable a resource to risk on hauling commercial satellites. Expendable rockets could do that job as well, and if they failed or blew up, the loss was only financial.

"The question is whether we should be sending up things on a manned shuttle when we could have done it without it," Feynman said. The physicist did not win support from the astronauts on this tack. Crippen argued that using men to deliver satellites to orbit "is a viable capacity." It should be used, he said. Weitz asserted that the system in the past had been using 110 to 120 percent of its resources. The deficiency in the flight program, he said, was a serious lack of spare parts, and that required frequent cannibalization of other orbiters to outfit the one scheduled to fly. The problem was spare parts, not cargo.

It was Henry Hartsfield who proposed an independent safety panel for the shuttle transportation system, a panel that would cover all levels of management and provide a clear channel to management on all problems.

Chairman Rogers, nodding his head in agreement, said the commission was coming to that conclusion. Hartsfield then summed up the general problem of flight safety as he saw it.

The space shuttle transportation system is certainly not an operational system in the traditional sense, and it can't be. It will never be routine, and there will always be risks associated with it, flying in space. We will fix all these problems and go fly again, but it still is going to be a risky business, and I think everyone should remember that.

The bottom line is, though, that this doggone vehicle is probably the most magnificent and fantastic machine I've ever seen, and it is something that we ought to be proud of. It has capabilities that are totally unmatched anywhere in the world. There's nobody that

has a machine like this, and if we use it properly I think it can do a great deal for this country.

Commenting on the risk, Feynman asserted that the problem was communication. That would be fixed by a safety panel with an astronaut on it, "because then you will be fully aware of all the things that are unsafe," he said. He had read all the flight readiness reviews and noted that they

agonize whether they can go even though they had some blow-by in the seal or they had a cracked blade in the pump of one of the engines; whether they go the next time or this time, and then it flies and nothing happens.

Then it is suggested, therefore, that the risk is no longer so high for the next flight. We can lower our standards a little bit because we got away with it last time. If you watch the criteria of how much blow-by you're going to accept or how many cracks or how the thing goes between cracks, you will find that the time is always decreasing, and an argument is always given that the last time it worked.

It is a kind of Russian roulette. You got away with it and it was a risk. You got away with it. But it shouldn't be done over and over again like that. When I look at the reviews, I find the perpetual movement heading for trouble.

Did Hartsfield mean a safety review board that would ride herd on whatever difficulty was presented at a review and try to get it fixed before the next flight? Dr. Feynman asked. He said he thought that was what was missing in the system.

Hartsfield agreed. That was his conception of the way the safety board would work, he said.

Landings and Brakes

John Young then cited three safety problems: landing at Kennedy Space Center in uncertain weather and

February 11, 1984. *Challenger* makes the first landing on the Kennedy Space Center's 15,000-foot runway at 7:16 A.M. EST on mission 41-B after 7 days, 23 hours in orbit. Vance Brand commanded the mission and Ronald E. McNair flew as mission specialist. The Vehicle Assembly Building is visible, *upper right. Upper center,* a chase plane monitors the landing. (NASA)

crosswinds; the orbiter's braking system, which he said was deficient in energy; and the nosewheel steering system, also deficient from the astronaut's view.[4]

A microwave landing system had been installed to assist the pilots in landing the orbiter, but none of them

[4] When *Discovery* landed at KSC runway 33 on April 19, 1985, one of its main wheel tires blew out and another was shredded when the wheels locked as the orbiter was braked to a stop. The Aerospace Safety Advisory Panel, an agency monitoring NASA operations, had warned in 1983 that the landing gear, tires, and brakes on the orbiters were marginal and were a possible hazard to the shuttle, according to *Science* (March 3, 1985), vol. 228.

favored it. Young said: "We're looking for ceilings with a microwave landing system in excess of 8,000 feet so that the crewmen can make the proper corrections in case those things are not working just properly. We're looking at crosswinds not very high because right now with the vehicle we have, we have a system that is single-string [no back-up] and nosewheel steering. There are numerous failures that can cause you to be no-string and nosewheel steering."

In addition, the energy-limited brake system was difficult to use with precision, he said, and "in fact

November 16, 1984. *Discovery* lands at 6:59 A.M. EST on the Kennedy Space Center's runway 15 after eight days in orbit on mission 51-A, Rick Hauck commanding. Five of the 24 missions landed here. (NASA)

we're finding out we don't really have a good technique for applying the brakes. On one landing we were told that we put the brakes on too long at too high an airspeed and kept them on too long. Then, the next landing, that was perfect as far as we were concerned . . . we were told we put the brakes on too short and kept them on, then put them on too hard.

"We don't believe the pilots should be able to break the brakes. That is the sort of— what has been happening to us when you land at Kennedy with the tires that we have because of the runway [it is hard on tires]. We are pushing to make our nosewheel steering more than single-string so that failures can't take the nosewheel system down."

Although he was a "Florida boy," Young said that he had a poor opinion of the Kennedy landing facility. True, it was 15,000 feet long and 300 feet wide. But,

he explained, the trouble was that the commitment to land there had to be made about halfway around the planet, nearly 90 minutes ahead of touchdown, and in that interval, the weather at the landing facility could change dramatically, for the worse.

The surface is very rough in a high crosswind, he went on. At the time the pilot executes his de-orbit maneuver,[5] the winds at Kennedy may be light, but when he makes his final approach an hour and a half later, crosswinds above the 10-knot limit may have sprung up or it may be raining, storming, or drowned in fog.

In a high crosswind, Young said, the cords tend to be scraped off the tires. The runway is surrounded by a

[5] For the de-orbit maneuver, usually done over the Indian Ocean, the commander fires the orbital maneuvering engines in the direction of flight to break out of orbit.

moat, and after heavy rains the water is near the runway surface. The Kennedy landing facility does not meet Air Force runway standards, he said.[6]

Young said that he and his colleagues preferred to land on the dry lake bed complex at Edwards Air Force Base in the Mojave Desert, with a runway 20 miles long and 7 miles wide. Or they preferred the alternate site at White Sands, New Mexico, where two intersecting runways provide the equivalent of 29,000 feet with a width of 900 feet.

It would be more prudent and safer for the program to land the shuttle at these runway complexes, he said. He admitted that "we spent a lot of money on that place [Kennedy] and we did a lot to it, but if we ever run off the runway at Kennedy the repair bill is going to be probably enough to build five or six more runways there at Kennedy."

Young concluded: "So we think the lake bed complex would be much better for the overall good of the space program. That is kind of what we recommend."

Chairman Rogers asked about the effect of rain at landing. Young said 10 to 15 seconds of light to moderate rain on the orbiter tiles (the heat shielding) would require repairs that would greatly extend the time to prepare the orbiter for the next mission. The silicate tiles are brittle and readily eroded by rain and hail. However, the runway is well grooved and might be safe (from skidding or sliding) when wet, he said.

The chairman asked how effective weather prediction has been at Kennedy, especially advance warning of rain squalls.

"One day we took Mr. Walt Williams [NASA consultant] out to give him a flight in this shuttle training aircraft," Young said. "There was one little thunderstorm sitting 13 miles off the end of the runway. Thirty minutes later there was a squall line across both ends of the runway. Before we went out, we checked with weather; they said there wasn't going to be any."

That type of quick weather development is not unusual at the space center, Young added, and it wouldn't affect a regular aircraft. But in addition to damaging the tiles, rain increases drag on the orbiter (which is a glider in the atmosphere), and a sudden increase in wind could prevent the glider from reaching the runway.

Rogers: What choice do you have once the decision is made to land at Kennedy and you have an hour and a half to go? Is there anything you can do . . . if the weather changes?

Young: Once you have been given the go for de-orbit, if you lose communication or if you don't have communications right up to the time you de-orbit— we waved off Crippen one time three minutes prior to de-orbit on 41-C [Challenger]. They reported to us that the weather was going to be clear at the time of landing. At the time of landing, there was 11,500-foot rain showers over the end of the runway. . . . I think we were about three minutes away from having Crip land in some pretty interesting rain showers.

Rogers: [repeating]: Once de-orbit occurs, is there any option left?

Young: No, sir. It is not like an airplane where any time you go somewhere in weather you always have an alternate. You are committed [in the orbiter] to land at one end of the runway or the other end. You can swap runways from about Mach 6, which is 12 minutes prior to landing, but that is the extent of your capability in terms of going to an alternate.

Commissioner Hotz: John, do you have any cross-range alternatives?[7]

Young: No, sir. We have talked about that, but the problem you get into there is that's even more harmful to the program. Suppose you did have a cross-range alternate and you were going to land at Orlando,[8] but then you're looking at a long time to get your machine back to Kennedy. What do you do? Close the Bee-Line [a toll expressway] and tow it? That would really be a tremendous problem

---

[6] It was the first time that complaint had ever been made public. There were complaints that alligators, protected as an endangered species, liked to bask on the runway.

[7] Cross-range is a deviation right or left of the flight path.

[8] Presumably, Orlando International Airport, with all traffic staying clear.

Neil A. Armstrong, who flew Apollo 11 to the Moon, considers orbiter flight problems as current astronauts air their views on flight safety to a national commission for the first time in the history of NASA. Armstrong served as vice chairman of the Rogers commission. (NASA)

to do that. You would probably have to chop up some underpasses and stuff.

Hotz: But it would be a safety alternative?

Young: It could be, but I'm not sure. Would it be worth the risk? I think you would be better off flying it off into Edwards and bringing it back [on the Boeing 747 transport aircraft] in four days or so.

Should the orbiter not land at Kennedy at all except in an emergency? Rogers asked.

Young pointed out that Kennedy was the only place the shuttle could land in an abort to the launch site. Otherwise, said the chief of the astronaut office, it was better to land on the dry lake bed in California.

Hartsfield agreed. "I for one began very much as he did, thinking that as soon as we get things squared away we ought to land at the Cape. I have since changed my mind based primarily on the fact that we have got a lot of problems with the brakes and the nosewheel steering. . . . we are pushing the state of the art on these brakes and so it is not an easy solution, but then the weather factor is certainly something to consider."

He recalled the first flight to land at Kennedy—STS 41-B *(Challenger)*, February 11, 1984 with Vance D. Brand commanding. ". . . If you look at that, it is a little frightening. You see patches of ground fog all over the place and pulling streamers off the wingtips as he [Brand] makes his final flare to land. And that wasn't predicted."

While Kennedy could be made safer, "you're never going to overcome the weather unpredictability," Crippen said. George Abbey, speaking as director of flight crew operations, seconded these views. "We've tried to get into Kennedy a number of times, and we have proven that we couldn't predict the weather. It has changed on us very rapidly. Crippen was waved off twice. In December we had three wave-offs."[9]

Commissioner Wheelon observed that if the orbiter is to land routinely at Edwards, using Kennedy only in emergency, the Boeing 747 carrier aircraft that hauls the orbiter back to the launch site in Florida becomes "a single-point failure" in the shuttle system. He suggested that purchase of a second 747 and its modification to carry the orbiter be considered.

Abbey agreed. He said he hoped it could be funded in the next NASA budget.

Autoland

Since 1976, NASA had been evaluating a Microwave Scanning Beam Landing System (MSBLS) that was capable of landing the orbiter automatically in ceiling zero visibility. Commercially, the system was known as

Autoland. Systems like it were in use by Navy aircraft carriers and by the Swedish and Finnish air forces.

Anyone who has seen the orbiter land is aware that it comes in very fast on final approach, much faster than an airplane. Its glide path on final approach begins at a steep angle. The orbiter is flared—leveled off—to reduce velocity to touchdown. To fly such a path in an unpowered, heavy glider automatically requires a computer-controlled autopilot and a ground signaling system that tells the computer where the vehicle is in relation to the touchdown point every microsecond of the descent.

This information is provided by the microwave scanning beam that sweeps across the landing sector. Pulses from a transmitter on the ground are coded to show the angle of the beam at each instant of its sweep. A receiver on board the orbiter picks up these pulses and decodes them to determine the orbiter's track.

With the data received, the orbiter's general-purpose computers compare the location of the vehicle, as signaled by the microwave beam, with the programmed descent path at every instant. If there is a discrepancy, the computer moves the orbiter's aerosurfaces to correct it.

On the afternoon of November 14, 1981, as Joseph H. Engle and Richard Truly were bringing *Columbia* in for a landing at Edwards Air Force Base, Mission Control in Houston gave them a "go" to switch to the Autoland system. It was a test. *Columbia* was slightly below the programmed glide slope at that moment.

Engle, commanding, released control to the Autoland system, and the microwave beam coming up the ground told *Columbia*'s computers to correct the flight path. Immediately, the computers reacted to the beam and pitched up *Columbia* to the planned glide slope.

The system worked that time, but it produced a side effect. The pitch-up, though brief, slowed the descent and caused *Columbia* to touch down short of the precise point marked on the runway.[10] On *Columbia*'s

---

[9] He referred to *Columbia* 61-C, originally scheduled for launch on December 18, 1985 but not launched until January 12, 1986. It was waved off Kennedy three times and landed in California. When a conventional aircraft is waved off, it may circle the airport until conditions allow landing. When the shuttle is waved off, it must circle the Earth.

[10] See Richard S. Lewis, *The Voyages of Columbia* (New York: Columbia University Press, 1984).

fourth flight, Mission Control told the crew to use Autoland on the approach to Edwards down to 2,500 feet. Then the commander, Thomas K. Mattingly, took control and landed manually.

The astronauts clearly were not happy with any automatic landing system and preferred to land the orbiter manually. Although Autoland would improve the crew's ability to fly the orbiter through low ceilings and poor visibility, these conditions usually implied landing the vehicle in the rain, Young commented.

"I suggest that you probably don't want to land the orbiter in those kinds of conditions for any reason," he said. "The Autoland system is dependent on many sensors, and it is always one failure away from not working at all. . . . we don't view Autoland as a practical solution to any problem in the space shuttle program."

It was Young's contention that shuttle crews cannot interact successfully with the Autoland system at low altitude. One reason is that the orbiter exhibits responses to the control stick that are the opposite of what the pilot intends, he said. Consequently, the pilot's natural reflex action to correct an emergency may be wrong.

Young explained: "we have these great big elevons on the back, and we have a little bitty attitude control stick, and it can move those big elevons an awful long way awful fast. And when that happens, exciting things happen.

"We actually train our pilots not to use the attitude control stick very much. They set up a trajectory, and the closer they get to the ground, if they are set up right, the less they move that control stick. And the orbiter exhibits what we call reverse attitude responses. And when it does that, a natural reflex takeover pilot action would be the exact wrong thing to do.

"For example, if you've got a sudden nosedown pitchover when you are doing Autoland, the reflex action to take over would be to pull back on the stick, to pull the nose up, and what it would do, it would raise those elevons, drop lift on the wing, and drive the wheels right into the ground at a high sink rate.

"And then the only way that people do operate on an Autoland touchdown machine safely is with a go-around capability provided by throttles. That is how you get yourself out of all kinds of jams with airliners, and needless to say the hundreds of approaches to touchdown that airliners must make to get FAA certified to Autoland we will never be able to do in an orbiter."

Young contended further that automatic approaches are not a successful way to fly the orbiter when the pilot has to do the landing. The Navy requires two independent autoland monitoring systems before they can do automatic landings, he said.

"At present, with the shuttle Autoland system, we have no independent monitoring system, and no one really knows how to implement one independent monitoring system for successful low-altitude crew takeovers."

Young conceded that automatic approaches were proposed in conditions where clouds or reduced visibility introduce more risk in the human-controlled approach, "but we feel that with only three orbiters left right now, that lower weather minimums is about the opposite of what the space shuttle program ought to want to do for a long time."

Weitz described safety problems at Dakar, Senegal, the shuttle's main transatlantic landing site. Visibility problems occurred frequently there. Braking and nose-wheel steering problems made it something of a gamble that a heavily laden orbiter could be stopped on a short runway. Night landings were difficult. Dakar often has poor visibility caused by dust and sand blowing off the Sahara and obscuring the runway, Weitz said.

Another safety issue was night landings. The orbiter has landed twice in darkness at Edwards. Weitz said the problem with night landings was the loss of visual cues the pilot needs. If there is a crosswind, it is much more difficult to perceive drift across the runway at night than in daylight.

Commissioner Neil Armstrong asked about Casablanca, Morocco as an alternate transatlantic abort landing site. Weitz explained that Casablanca was "on order." Inasmuch as the shuttle is launched at 28.5° north latitude, it would have to change its orbital inclination to 31° or 31.5° to land at Casablanca at the higher latitude. Such a maneuver would impose a performance penalty of 2,000 to 3,000 pounds of fuel, Weitz estimated. "In those cases where we have performance, we would like to consider shaping to go do it, which requires more resources."

Commissioner Kutyna asked Young: "If you go into Dakar on a TAL [transatlantic landing] with some of the heavyweight payloads that we have planned now and the brakes you've got now, can you stop?"

"It depends upon what brake energy margin and what kind of a headwind or tailwind you've got," said Young. "You have to put the brakes on with 5,000 feet to go in a no-wind situation when you are doing 165 knots. You are putting those brakes on so darned fast that you use energy up just like nothing flat. If they don't hold up, if they don't give you the full brake amount, which is 55 million foot-pounds . . . I know you would be in real trouble on stopping."

He added that the brakes had failed at 44 million foot-pounds and also at 34 million pounds. "We still don't understand the reason for that failure."

Kutyna: And what's at the end of the runway?
Young: Well, there's a blockhouse and a cliff.

More of an Art than a Science

There is an ongoing program for improving the brakes, and there has been for some time, said Crippen. "It is not so much developing new technology, but we have discovered that the making of aircraft brakes is more of an art than it is a science. And especially on airplanes, you have the opportunity to take them out, do ground runs, and make adjustments for them, and we really don't have that kind of a facility available to us. Consequently, we are asking a new lab at Wright Pat [Wright Patterson Air Force Base, Ohio] to help us

do some of those kinds of tests so that we can improve the brakes."

Henry Hartsfield amplified some of John Young's comments on orbiter handling problems. Although every astronaut who has flown the orbiter praises it as a magnificent flying machine, it does have some very strange characteristics when you get it down close to the runway, he said.

"Neil will understand this, I know," said Hartsfield. "But the pilot is located at or slightly aft of the apparent center of rotation, which is kind of a bad place for the pilot to be. It is just the nature of the airplane because of the large percentage of the wing area that is in the elevons themselves. And the end result is what John talks about, when the pilot makes an input, he doesn't get any physical feedback that he has done something. And so you have to fly the orbiter pretty much open loop. You make small inputs and then wait and see what happens, and that is not a natural pilot instinct. So we spend a lot of time training the pilots in the training airplanes."

The thrust of the astronauts' detailed recital of orbiter piloting problems was to support a plea for additional training aircraft as a safety measure. The recital itself was unique. For the first time in the American space program, astronauts were talking freely and publicly about the problems that concerned them. For the most part, they were talking to fellow professionals who understood what they were saying; and it should be said that the members of this jury who were not aerospace professionals could appreciate it as well—one a former secretary of state and former Attorney General serving as chairman, the others a theoretical physicist, a Washington attorney, and a retired aerospace magazine editor.

This session, in which the views from the flight deck were aired by television to a national audience, was a far cry from congressional space committee hearings, where protocol and agency discipline tend to inhibit criticism. From a journalist's viewpoint, there was more insight than headlines in this commission hearing, yet it was an epochal event. Largely at the instigation of the chairman, it evolved into a rap session on the real problems of flying the space shuttle. Once the astronaut witnesses realized they had an attentive and sympathetic jury, they issued a candid diagnosis of the faults of the system.

Hartsfield said there is a problem for the pilot in sensing what is happening around the machine at touchdown because of the pilot's eye height, 33 feet above the ground. That is in the same ball park as the 747—and its pilots have perception problems, too, he said.

"I have talked to these guys on how they land those things," Hartsfield said, "and they pretty much set up a steady rate of descent and just fly them on. Our problem is aggravated by the fact that we are decelerating rather rapidly and losing over five knots a second. We are just a glider, and so we have to set it up and get it on the ground. As John says, we learn not to make inputs close to the ground because anything you do is wrong.

"For example, at the sink rate, if you sense the sink rate is a little bit too high, the instinct is to put in a little backstick. As soon as you do that, you move the elevons. One degree of elevon at 200 knots is worth 6,000 pounds of lift, so you move them 3 or 4 degrees and you've really dumped a lot of lift, and the thing really drops in on you. So it takes a lot of time to learn this technique."

To outsiders, the process of bringing the orbiter in for a landing had appeared to be as routine as landing any large aircraft. But it was quite different, as the men who fly the orbiter described it. Hartsfield had assumed the role of a witness appealing directly to a jury of his peers for help in getting adequate funding for pilot training. It was the first instance that funds for such a fundamental safety program as pilot training for space flight had been presented publicly as a concern. It was the right place for such an appeal, for the commission's safety recommendations would go directly to the President as well as to Congress, and through this commission the astronauts had the President's ear.

The situation was: there were two kinds of devices essential for the pilot training program, and both were wearing out. One was the modified Gulfstream II, the shuttle training airplane. "That is where we really learn to fly the orbiter and learn to land it, and we can't do this in a simulator," Hartsfield said. "Well, what I'm leading up to is we are beginning to get in a lifetime problem with those airplanes, and that is starting to worry us."

By the time a pilot becomes a commander and lands the orbiter for the first time, he will have made about 900 approaches with that training airplane. Hartsfield went on: to support 12 shuttle flights a year, the rate toward which NASA was moving at the time of the *Challenger* accident, about 1,400 hours of training a year were required in the system's three training airplanes. Two of them now have more than 5,000 hours on them, about 1,200 more than the predicted fatigue life of some components, such as basic wing structure.

"Our third STA [shuttle training aircraft] doesn't have that much time on it, but we need another airplane. Otherwise, I'm afraid that we won't be able to support the flight rate."

The other training item causing concern, he said, is the shuttle simulator. Johnson Space Center has two, one with a full cockpit mock-up that doesn't move and the other with a moving base. At the time of the accident, crews were training 148 hours a week on the simulators, he said, but 180 hours a week are required to support a flight program of 12 to 14 missions a year.

"In other words," said Hartsfield, "at the time that we had the accident, we were already facing up to a crunch on crew training." That problem is complicated by obsolescence, he added. The simulators reflect 1970 technology, and their computers are no longer being manufactured. He said that the vendor warned "he won't make parts for them in three more years."

Although the budget was projected to upgrade the simulators, there are other problems, Hartsfield continued. "We have a pretty lousy visual system. We know that, and we're trying to get something on line to improve it. We have a large number of models in the simulator that are not what they ought to be. As an example, the main engine model. The interaction between the crew and the simulation of the shuttle main engines is so bad that in some cases we get negative training, and it is one of our priority items to fix. . . . over 600 discrepancies against the simulators got us worried about maintaining our training. And what we really needed, I think, is another SMS [shuttle mission simulator] in order to support a flight rate of a dozen or more and do it comfortably."

Commissioner David C. Acheson, attorney, asked how the safety function is organized in NASA.

George Abbey responded. "Both John and I were very much involved in the activities following the Apollo fire in 1967, and at that point there was considerable strengthening of the reliability and quality assurance and safety organization within NASA. We established a safety office within the Office of Space Flight. We also had an independent reliability and quality assurance and safety organization at each of the centers reporting to the directors."

During the Apollo program, emphasis was placed on identifying all the anomalies that appeared—"right up to the top level, up through NASA headquarters, and they were reported and acted upon and satisfactorily resolved."

Abbey said that he believed the same attitude existed within the shuttle program on the flights before 51-L, but with the attempt to fly more frequently, "I think we probably need to look at a better check-and-balance of the system." He said he and Young had discussed establishing a more independent safety organization within NASA and its centers.

"I have this feeling," said Young, "that the very biggest problem that must be solved before the space shuttle flies again is that one of communication . . . with respect to the early identification and proper appreciation of program-wide safety issues."

Before the accident, he said, many safety issues within the system were not being worked, "I was told mainly because we didn't have the money to deal with them." That was "a worrisome condition."

The shuttle has certain risks associated with its normal operation, "if you can call what it does normal," he said. Everyone was working hard to make it operational. "I think once you get it on orbit and start doing what people are supposed to do in space, it really is what you can call an operational system. I wonder though sometimes why the space shuttle is inherently risky; why we should accept additional, affordable risk in order to meet launch schedules . . . and we do that sometimes, or to reduce operating costs, and that has been proposed, or to fly unsafe payloads. And I think sometimes that happens. We just can't afford to have another accident.

"One of the problems we have is to get a communications link and properly define those risks. We need a foolproof way to surface to the top and correct safety issues early so that we can prevent another accident.

"There is a great bunch of engineering people at NASA. I guarantee . . . that all the working troops right this minute know exactly what all the space shuttle issues are right this minute.

"But what we have to worry about is five years from now, when Joe Engineer comes in to his boss and he says, 'Hey, how about this data here that shows the framistam keeps breaking and it's going to blow the side off the orbiter. And his boss says, 'That hasn't failed in 60 flights, get out of my office.'

"And so here's something that's bad that could happen . . . and it doesn't get through the system because his boss has a million things on his mind. The way I think you can prevent that, one way, would be to get . . . an agency-wide safety organization, similar to those in many airplane programs. The safety people would be independent of the cost and schedule concerns of their branches or divisions or directorates or centers.

"Unless we take very positive steps to open safety communications and to identify and fix early on safety problems, we are asking for another shuttle accident."

9. Findings

The report of the Presidential Commission on the Space Shuttle *Challenger* Accident is one of the most critical reviews of an agency's operations in the history of American technology. It charges management oversight or failure at every level of the National Aeronautics and Space Administration's stewardship of the National Space Transportation System, including headquarters—Level 1.

The report, delivered to President Reagan on June 6, 1986, concluded "that the cause of the Challenger accident was the failure of the pressure seal in the aft field joint of the right solid rocket motor."

It added: "The failure was due to a faulty design unacceptably sensitive to a number of factors. These factors were the effects of temperature, physical dimensions, the character of materials, the effects of reusability, processing and the reaction of the joint to dynamic loading."

Failure of the joint, the report said, began with "decisions made in the design of the joint and in the failure by both Thiokol and NASA's solid rocket booster project office to understand and respond to facts obtained during testing."

The report lays the blame for the accident on both the space agency and its solid rocket motor contractor, Morton Thiokol, Inc., Brigham City, Utah. Neither responded adequately to internal warnings about the faulty seal design, the commission concluded. Nor did either make a "timely effort to develop and verify a new seal after the initial design was shown to be deficient."

"Neither organization developed a solution to the unexpected occurrences of O-ring erosion and blow-by, even though this problem was experienced frequently during shuttle flight history," the report said. "Instead, Thiokol and NASA management came to accept erosion and blow-by as unavoidable and an acceptable flight risk."

Perhaps the core of the commission's criticism was its finding that "prior to the accident, neither NASA nor Thiokol fully understood the mechanism by which the joint sealing action took place." The report makes it clear that this deficiency led to the worst accident of the space age, causing a catastrophic collapse of the National Space Transportation System and grounding the shuttle for at least two years.

In addition to the failure to understand the O-ring seal operation, the commission found that the joint test and certification program by NASA and Thiokol was inadequate because "There was no requirement to configure the qualification test motor as it would be in flight, and the motors were static tested in a horizontal position, not in the vertical flight position."

Moreover, the commission found, NASA and Thiokol accepted escalating risk apparently because they "got away with it last time," as Commissioner Feynman put it. The report quoted his observation that decision-making was "a kind of Russian roulette. . . ."

Despite a history of persistent O-ring erosion and blow-by in the field joints of the motors, NASA's system for tracking anomalies for flight readiness reviews failed, the report said, and flight was still permitted. It failed again in the "strange sequence of six consecutive launch constraint waivers prior to 51-L, permitting it [the shuttle] to fly without any record of a waiver or even an explicit constraint."

The report stated that the O-ring erosion history presented to Level 1 at NASA headquarters in August 1985 was sufficiently detailed to require corrective action prior to the next flight. [But none was taken.]

The commission noted that a careful analysis of the flight history of O-ring performance would have revealed the correlation of O-ring damage and low temperature, but neither NASA or Thiokol carried out such an analysis. ". . . Consequently they were unprepared to properly evaluate the risks of launching the 51-L mission in conditions more extreme than they had encountered before."

What the Task Force Reported

The commission's findings were based on information obtained from three sources: the Data and Design Analysis Task Force, composed of NASA personnel and headed by Rear Admiral Richard H. Truly, associate administrator for space flight; a series of investigative hearings; and on-site interviews and investigations by members of the commission and its investigators.

The task force report essentially relates the sequence of events leading to the accident and reconstructs the manner in which it happened. This is a narrative derived from telemetry data from *Challenger* through 73 seconds of flight, photographic and video images, and other ground-based observations. In addition, the task force derived information from the reconstruction of a mathematical model of the shuttle's response to upper-atmosphere winds during launch and ascent.

The task force reported that the mission, launched at 11:38 A.M. January 28, 1986, proceeded normally through the start-up of the orbiter engines and the ignition of the solid rocket boosters. The first anomaly seen was a small puff of smoke between the right booster

and the external tank in a region near the booster's aft field joint.

The smoke appeared to persist for a period between 0.678 and 3.375 seconds. From this point onward to approximately 59 seconds all systems appeared to perform within their design boundaries, the report said. However, there were significant vehicle attitude excursions and responses by the high thrust vector control (steering) system beginning at 37 seconds. This activity, created by upper-atmosphere wind gusts and planned maneuvers, persisted through 59 seconds, when the vehicle was heavily loaded by dynamic pressure.

At 58.788 seconds, a flame appeared, flickering, from the general region where the puff of smoke was observed near liftoff. The flame grew in intensity and size until 59.262 seconds when it was masked by a large plume of smoke. This event occurred about five seconds after the solid rocket motors went through a reduction in chamber pressure (by design) at 54 seconds, and the motor pressure was again rising.[1]

At 60.004 seconds, the right solid rocket motor's internal pressure began diverging from that of the left motor. This correlated with a right motor combustion gas leak. At 61 seconds, the well-defined plume was seen to be deflected, indicating that the hot gas had reached the external tank. Photographic analysis indicated that the plume breached the tank and produced a liquid hydrogen leak at 64.660 seconds. The leak was confirmed two seconds later when the tank could no longer maintain its normal repressurization rate. At 72.6 seconds, tank pressure could no longer be maintained, indicating the leak path had increased and was growing rapidly.

At 72.2 seconds, the guidance system showed that the right booster motion was not the same as that of the orbiter and the left booster, indicating that the lower external-tank-to-booster attachment strut was severed or

---

[1] The motors were configured to reduce thrust at this point in the ascent, at the same time the orbiter engines were throttled back, to reduce buffeting as the shuttle passed through the region of Max Q, maximum dynamic pressure.

was pulled loose from the tank. Large steering commands and system responses were then observed, and at approximately 73 seconds both liquid hydrogen and liquid oxygen pressures to the orbiter main engines showed a significant drop. This was followed at 73.124 seconds by the appearance of a circumferential white pattern around the aft bulkhead of the external tank, suggesting a structural failure of the hydrogen tank.

At 73.137 seconds vapor was observed at the intertank region, indicative of the liquid oxygen tank failing. The failure of the liquid oxygen tank can be attributed to abnormal loads induced by either or both the right booster action at the forward attachment point or the propulsive forces created by the structural failure of the liquid hydrogen tank's aft bulkhead.

Liquid oxygen was then seen streaming along the tank. At 73.191 seconds a flash was seen between the tank and the orbiter, and that was immediately followed by a total-vehicle structural breakup explosion at 73.213. Intermittent telemetry data were obtained until 73.7 seconds, at which time the main engines were noted responding normally to reduced propellant pressures from the ruptured tank. Both boosters then escaped the vehicle breakup propulsively and continued to fly erratically until destroyed by range safety command at 110.25 seconds.

The Anomalies

The task force examined prelaunch activities that might have had a bearing on the accident and the effect of the weather on the functioning of the solid rocket motors. Launch site records showed that although the right booster aft field joint was mated by approved procedures, "significant out-of-round conditions existed at mating."

In order to mate the aft-center segment to the aft segment, the report said, a special tool was used to adjust the diametrical shape of the aft-center segment.

Even with the use of the tool, the relative diameters of the two segments allowed the tang end of the joint to interfere with the inner leg of the joint clevis and also with the O-rings.

The area of maximum interference around the circumference of the joint was located in the area of 120 to 300 degrees. Across the diameter of this circumferential area, there was a difference of 0.393 inches at mating between the outer diameter of the tang and the inner diameter of the clevis outer leg—that is, the tang diameter was less than the clevis diameter. The task force said that this condition indicated the existence of maximum compression or squeeze on the O-rings at the 300 degree location where the joint leak occurred.[2] Tests conducted during the investigation indicated [among other effects] that maximum squeeze is a significant factor in the functioning of the O-rings because it compresses the O-ring against its groove walls, and this prohibits its actuation by engine pressure to seal the joint.

The finding that the location of the field joint combustion gas leak coincided with the approximate location of interference between the joint tang and clevis at assembly raised a question about mating procedures when the motor segments were put together in the Vehicle Assembly Building. The task team report stated that "this mating interference can exist using approved assembly procedures which for 51-L were implemented properly." However, it added, mating conditions at this location could result in maximum squeeze on the O-ring.

The task team also considered the effect of cold weather. It noted that ambient temperature at ground level at the launch pad during the 24 hours before launch reached a low of 24° F at 7 A.M. EST, while at launch time, 11:38 A.M., it was 36° F. This translated to a right solid rocket booster temperature of 28° F at launch time, the report said. The right booster joints had

---

[2] The 300 degree circumferential location was approximately where the booster was attached to the external tank.

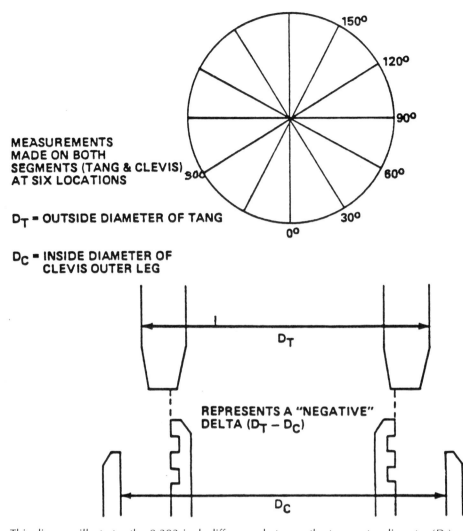

MEASUREMENTS MADE ON BOTH SEGMENTS (TANG & CLEVIS) AT SIX LOCATIONS

D_T = OUTSIDE DIAMETER OF TANG

D_C = INSIDE DIAMETER OF CLEVIS OUTER LEG

REPRESENTS A "NEGATIVE" DELTA ($D_T - D_C$)

This diagram illustrates the 0.393-inch difference between the tang outer diameter (D_T) and the inner diameter of the clevis outer leg that existed at mating. The tang diameter was thus less than the clevis diameter. This condition indicates that maximum compression, or squeeze, was exerted on the O-rings at the 300° location where the joint leak occurred. It would have interfered with their performance in sealing the joint. (Report of the Accident Analysis Team, *Commission Report*, vol. 2, p. L-4)

larger circumferential temperature gradients than those on the left booster, the commission stated.[3]

It recalled that the flight with the next-coldest solid rocket motor temperature—53° F—was 51-C (when significant seal damage occurred). The low overnight temperatures before the launch of 51-L required the pad crews to allow small, continuous water flow in the pipes of the pad fire-extinguishing (FIREX) system to prevent

them from freezing. This flow accounted for the ice on the pad, as mentioned previously.

Turbulence in the atmosphere during *Challenger*'s ascent also was treated as an anomaly. At 37 seconds after liftoff, 51-L encountered the first of several tubulent wind conditions, which lasted until 62 seconds. The effect was "relatively large fluctuations in forces applied to the vehicle." It caused rapid changes in the vehicle's attitude that were immediately sensed and corrected by the guidance navigation and control system.

Such corrections in flight have little significance,

[3] *Report of the Presidential Commission on the Space Shuttle Challenger Accident* (Washington, D.C., 1986), vol. 1, p. 62.

the task team said, unless they induce unusual loading conditions or cause the control system to exceed its design limit. While that did not happen, the early yaw plane (left and right) gusts were significant, and at one time exceeded prior flight experience in the subsonic region.

Pitch plane (up and down) gusts were large also, and at two times, 55 and 68 seconds into the ascent, dynamic pressure (in terms of pounds per square foot) exceeded previous flight experience, the report said. However, the maximum forces in both pitch and yaw that *Challenger* encountered "were well with design limits," the report said.

Effects of Temperature

The calculated solid rocket motor joint temperature at launch (28° F plus or minus 5 degrees) had two effects with potential impact on the seal performance, the commission report said.[4] It degraded the resiliency of the O-rings and thus reduced their ability to seal the joint, and it created a potential for ice in the joints. "Consistent results from numerous O-ring tests have shown a resiliency degradation with reduced temperatures," the report said. "It is probable that water intrusion occurred in some if not all 51-L field joints." During the time that *Challenger* stood on the launch pad (37 days), it was exposed to approximately 7 inches of rain.

"Water had been observed in STS-9 [*Columbia*] joints during de-stacking operations following exposure to less rain than that experienced by 51-L," the report noted. "Water had drained from STS-9 when the pins were removed and one-half inch of water was in the bottom of the clevis."[5]

The report cited tests indicating that water in the joint will freeze under the environmental conditions

[4] Ibid.

[5] After being erected on the launch pad, STS-9 was taken down and de-mated. A suspect nozzle in the right booster was replaced after the booster was de-stacked.

experienced prior to the 51-L launch and could unseat the primary O-ring. Tests on a subscale dynamic fixture at Morton Thiokol showed that with ice present, the secondary O-ring failed to seal.

The task team looked at problems with the putty that had been debated for two years before the accident. The putty may act as a seal during the ignition transient (about 0.6 seconds) and may not allow motor pressure to seat the primary O-ring. If the putty then ruptures and allows motor pressure to reach the O-rings after joint rotation has opened a gap between tang and clevis, the potential exists for the O-rings not to seal, the task team found.

A motor pressure delay (by the putty) of 500 milliseconds (one-half second) would result in a flow path greater than 0.010 inches between the O-rings and the tang sealing surface, the team reported. Tests indicate that sealing capability is marginal for maximum squeeze conditions—0.004 inches initial gap at 50° F—with a pressure delay of 500 milliseconds.

The report then applied this reasoning to *Challenger*'s right solid rocket motor: "For the temperature and O-ring squeeze conditions which existed for several of the 51-L field joints, sealing was not achieved in these tests with simulated putty rupture times of 250 and 500 milliseconds."

The Joint Leaked the First Second After Ignition

The Data and Design Analysis Task Force then came to these conclusions, which served as the basis for the Presidential Commission's final report on the cause of the accident:

1. A combustion gas leak through the right solid rocket motor aft field joint weakened and/or penetrated the external tank hydrogen tank initiating structural breakup and loss of 51-L.

2. The evidence shows that no other STS 51-L

shuttle elements or the payload contributed to the cause of the right solid rocket motor aft field joint combustion gas leak.

3. The joint was observed leaking combustion gases *within the first second* after ignition. The leak became clearly evident at approximately 58 seconds into the flight. It is possible that the leak was continuous but unobservable or nonexistent in portions of the intervening period.

In either case it is possible that vehicle response to wind gust and planned maneuvers contributed to the leakage growth at the degraded joint.

4. The location of the right solid rocket motor aft field joint combustion gas leak coincided with the approximate location of maximum interference between the joint tang and inner clevis leg at assembly. This mating interference can exist using approved assembly procedures.

Mating conditions at this location could result in maximum squeeze on the O-rings.

Mating conditions there afforded the highest potential for undetected joint damage and for contaminant generation which could degrade performance.

5. The solid rocket motor joint has design deficiencies and may not seal properly with combined variable conditions existing on 51-1. Tests conducted during this investigation indicated that:

The joint sealing performance is significantly degraded as joint rotation increases the gap when O-rings are subjected to temperatures below 40–50 degrees F. and maximum squeeze conditions are in the range of 0.004 inches. The 51-L O-ring temperature was approximately 28 degrees F. and maximum squeeze was probable.

Maximum squeeze is a significant factor. . . . and is aggravated by the design leak test that places the primary O-ring in an unfavorable position making it more difficult to pressure actuate the O-ring to its proper sealing position.

Putty performance in the joint may not be as intended by the design. Tests indicate that humidity can delay timely actuation pressure to the O-rings. Testing and the destacking of the boosters for 61-G [intended for *Atlantis*] showed that several blow-hole gas paths

per joint can exist at motor ignition and can concentrate hot gas flow, causing O-rings to erode.[6]

Ice in the joint was cited as another factor inhibiting seal performance.

6. The joint that failed was not found to be unique. It was therefore necessary to modify the solid rocket motor joint design to preclude or eliminate the effects of all these factors and conditions acting singly or in combinations.

Crew Escape

The question of crew escape that the astronauts had discussed on April 3 with the commission was investigated in considerable detail by the task force. It reported that the National Space Transportation System (NSTS) had considered all known methods of providing the crew with a facility to escape from emergency situations during the first stage of the ascent.

Ejection seats and pressure suits were provided for the four-mission flight program in *Columbia*.[7] These, the task force acknowledged, provided only limited capability for commander and pilot to save themselves at launch. Other methods were studied for the "post flight test period," the task force reported, but were never implemented. The reasons: limited utility, technical complexity, lack of time to react to an emergency and to appropriate cues, the cost of an escape system, and the impact of installing it on flight schedules and mission objectives.

"Because of these factors," the report said, "NSTS adopted the philosophy that the reliability of first stage ascent must be assured through conservative design, testing and certification to preclude time critical failures that prohibit the continuation of flight through solid rocket booster burnout."

After burnout and the jettisoning of the boosters,

---

[6] It was found that prelaunch blowholes in the putty were created by the leak check that tested the seals under pressures up to 200 pounds per square inch.

[7] The seats were similar to those in the Lockheed F-12/SR-71 aircraft.

Shuttle Abort Regions

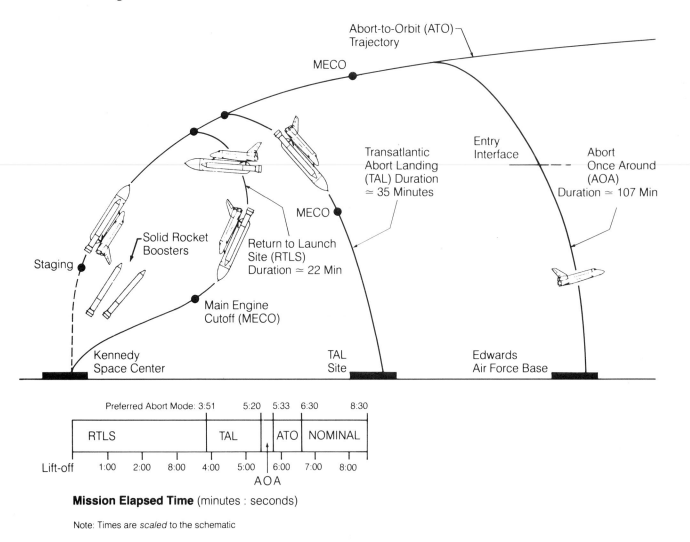

Mission Elapsed Time (minutes : seconds)

Note: Times are *scaled* to the schematic

Schematic shows options available to Space Shuttle crews for aborts in the event of power loss at various stages in the ascent to space.

This chart shows the abort options in case of main engine failure. No abort option is available until the solid rocket boosters burn out and drop after the first two minutes of the ascent. Up to 3 minutes, 51 seconds of the ascent, the crew can turn the orbiter around and fly back to the Kennedy Space Center landing site, dropping the external tank en route while over the ocean after main engine cut-off (MECO). The next option is to make an emergency landing at a site in Africa (at the time of the *Challenger* launch, it was Dakar, Senegal) up to 5 minutes, 20 seconds. The next abort option was once around the Earth with landing at Edwards Air Force Base, California. After 5 minutes and 33 seconds of powered ascent, the crew could abort to orbit if the engines failed and attempt to carry out at least part of this mission. (NASA) abort to emergency landing sites becomes possible, but that is related only to the performance of the orbiter main engines. Thus, from the beginning, the shuttle design eliminated launch escape with the rationale that the boosters could be made to work flawlessly.

During the shuttle development period, the task force related, the question of first-stage abort "was revisited many times by all levels of NASA management from 1973 to 1983 with no change in the philosophy that reliable first stage ascent must be assured."

The options available to the program today, the report went on, appear to be the same as those considered early in the program. There is the added complication that implementing major changes such as escape pods within the orbiter design would not be possible.

Ejection seats limit crew size and thereby the utility of the shuttle. Moreover, the report said, they provide very limited increase in survivability for the total range of potential first-stage failures. Bail-out options offer an escape path for the crew in the contingency abort cases that result in water ditching, but bail-out during the first stage of the ascent, while the boosters are firing, is impossible.

It is known to be possible to shut down a solid rocket motor before all its fuel is consumed, a tactic called thrust termination. The task team said that an escape pod with termination of booster thrust theoretically offers the widest range of crew escape options during the first stage of ascent, but it requires a system to detect an impending failure.

An escape pod can offer an opportunity for crew escape at all altitudes during first-stage contingencies if the pod is not damaged to the point that it cannot function. "The implications of implementation are a significant redesign of the orbiter and major loss of payload capacity," the task force said.

In its final report, the Rogers commission went considerably beyond the task team's pessimistic assessment of launch escape possibilities for the shuttle. It repeated the condition that the "pod" or "escape module," as the commission called it, is not hopelessly

damaged by the accident. It added that the module must be sufficiently far from the vehicle at the time of catastrophe that neither it nor its descent system is destroyed.

In addition to requiring significant redesign of the orbiter, the incorporation of the escape module would require some structural reinforcement, pyrotechnic devices to sever the module from the rest of the orbiter, modifications to sever connections that supply power and fluids, separation rockets, and a parachute system.

The commission report mentioned that an additional weight penalty would be imposed by the requirement of adding mass in the rear of the orbiter to compensate for the forward shift in the center of gravity created by the escape module. The increase of weight was estimated at about 30,000 pounds.

"This increase in weight would reduce payload capacity considerably, perhaps unacceptably," the report said. "There is no current estimate of the attendant cost."

The commission concluded that "an escape module does theoretically offer the widest range of crew escape options. The other two options, rocket extraction and bail out, are only practical during gliding flight."

In all cases, although the crew might be saved, the orbiter itself would be lost. The commission tended to view the escape module in a more positive way than did the task force which concluded its assessment of launch escape with this comment: "In summary, the NSTS must maintain and be successful with its previous first stage design philosophy of ensuring first stage reliability through design and certification to preclude failures."

Why this design philosophy has failed thus far in the shuttle program is illustrated by the following responses to questions by the commissioners:

Q: Why are there no specific O-ring and putty launch temperature criteria?

A: MTI [Morton Thiokol, Inc.] said there were no specific O-ring and putty temperature criteria in the contract . . . and that only thermal criteria for propellant mean temperature applied to solid rocket motor operation. MTI interpreted 31° to 99° F natural environment as a prelaunch environment, not a flight operations requirement. Marshall Space Flight Center failed to notice this omission in the design certification reviews.

Q: What analyses of blowholes in the putty have been carried out? What have been the results?

A: MTI stated that they had not modeled the putty blowhole mechanism. However, an empirically based model has been developed which characterizes the O-ring pressurization and erosion resulting from the putty blowhole. The model has validated and recent runs show that O-ring damage varies directly with cavity size and cavity heat loss and inversely with heat jet size.

The Development Team Report

The report of the task force's development and production team illustrates further why the presumption of first-stage reliability as a reason to fly without a launch escape system was unrealistic. The team found and concluded that:

All environmental specifications imposed on Morton Thiokol by NASA were not adequately verified by test, analysis or similarity. The field joints which are susceptible to environmental conditions were not qualification tested to the full range of the contractually required environments. This led to a lack of complete understanding of the joint design limits.

Prior to the STS 51-L accident, there was insufficient knowledge on the part of MTI and NASA relative to the performance characteristics of the putty used in the solid rocket motor field joints.

Prior to the STS 51-L accident there was a lack of understanding on the part of MTI and NASA of the joint operation as designed. There were insufficient development tests and analysis performed by MTI to understand the individual and/or combined effects of the putty performance, joint rotation and O-ring compression and resiliency on the pressure sealing integrity of the field joint.

The configuration of the qualification test article was not in all cases representative of the flight configuration; e.g., the qualification motor lacked the lift-off dynamic effects on the case joints and the putty preparation differed from that used on flight hardware.

In order to verify the adequacy of flight hardware . . . it is essential that the configuration and processes be as similar as practical. Deviations from standard practice can lead to a lack of understanding of the performance of the system.

For a period of time which included the processing of the 51-L O-rings there was an elimination of mandatory inspection points from the O-ring acceptance criteria at MTI. The secondary O-ring for the suspect 51-L field joint was processed during the time these mandatory inspection points were not in effect.

Although inspection records indicate a high probability that the 51-L O-rings were acceptable, and the inadvertent elimination of inspections has been corrected, the report said, "the possibility exists that other areas similar in criticality may be suspect if a complete assessment of all mandatory inspections is not performed."

The full-scale hot fire static-testing of the development qualification motors was performed with the motors in the horizontal position rather than in flight attitude. The team said this decision was based on "programmatics since the horizontal test facility was in place and the concern of being able to determine . . . the actual thrust produced by the motor." It was not obvious, the team added, that the effect of this deviation in test and flight configuration received sufficient attention.[8]

In regard to the reuse of the boosters after recovery from the sea, it was found that remeasurement of two used case segments indicated that both tang and clevis sealing surfaces had increased in diameter beyond anticipated design limits. The cause of this phenomenon

---

[8] Firing the motor in its flight configuration would have required the construction of a vertical test stand by Morton Thiokol or NASA.

appeared to be uncertain. The uncertainty must be resolved, the team said, before reuse of case segments is planned.

The Safety Program

The commission report expressed surprise that NASA's safety staff was never mentioned in the testimony. The question of flight safety had been relegated so far into the background of NASA activity since Apollo that the agency's safety program was ignored.

No witness related the approval or disapproval of the reliability engineers and none expressed the satisfaction or dissatisfaction of the quality assurance staff.

No one thought to invite a safety representative or a reliability and quality assurance engineer to the January 27, 1986 teleconference between Marshall and Thiokol.

Similarly, there was no representative of safety on the mission management team that made key decisions during the countdown on January 28, 1986.

The commission stated that it was concerned "about the symptoms it sees." It remarked: "The unrelenting pressure to meet the demands of an accelerating flight schedule might have been adequately handled by NASA if it had insisted upon the exactingly thorough procedures that were its hallmark during the Apollo program."

The report recalled that when Arnold Aldrich, shuttle program manager, appeared at its hearing on April 3, he related five different communication or organization failures affecting the launch decision, and four of them related to faults within the safety program. These including a lack of problem reporting requirements, inadequate trend analysis, misrepresentation of criticality, and lack of safety program involvement in critical decisions.

"A properly staffed, supported and robust safety organization might well have avoided these faults and thus eliminated the communication failures," the report said. "NASA has a safety program to ensure that the communication failures to which Mr. Aldrich referred do not occur. In the case of mission 51-L, that program fell short."

The overall responsibility for safety, reliability, and quality assurance in the national space program is vested in the chief engineer at NASA headquarters, but the commission found that the ability of this official to manage such a program is limited. His staff of 20 persons includes one who spends 25 percent of his time and another who spends 10 percent of his time on safety, reliability, and quality assurance, the commission reported.

Further, it found no independent, centralized, nor effective safety organization in NASA was monitoring problems. At the Johnson Space Center, Houston, there were a number of engineers working in safety, reliability, and quality assurance organizations, the report said, "but expertise on Marshall hardware is absent."

At the Kennedy Space Center, there were a "myriad" of people in safety, reliability, and quality assurance functions. In most cases, the commission reported, they report to supervisors responsible for shuttle processing. This practice fails to provide an independent role necessary for flight safety.

At the Marshall Space Flight Center, the director of reliability and quality assurance reports to the director of science and engineering, who oversees the development of shuttle hardware. This results in a lack of independence from the producer of hardware, the commission said.

Trends

In the commission's view, any integrated safety-reliability program in the space agency could not function because reporting of flight anomalies and hardware problems was so poor it failed to disclose dangerous trends. One of these was deterioration of the joint seals.

The commission noticed than an "ominous trend"

began in the booster program in January 1984 following STS-9, the flight on which *Columbia* carried the first Spacelab mission. Until then, there had been only one serious O-ring anomaly in a field joint (*Columbia* STS-2, November 12, 1981) in the first nine shuttle flights.

Beginning with the tenth flight, *Challenger* 41-B, through the twenty-fifth, *Challenger* 51-L, more than half of the missions experienced field joint O-ring blow-by or erosion.

The commission commented: "This striking change in performance should have been observed and perhaps traced to a root cause. No such trend analysis was conducted . . . the significance of the developing trend went unnoticed. The safety, reliability and quality assurance program exists to ensure that such trends are recognized when they occur."[9]

The commission noted that there had been changes in solid rocket booster processing procedures at the Kennedy Space Center, where the booster segments shipped from Utah were mated. The changes included discontinuance of O-ring inspections, the doubling of leak check stabilization pressure from 100 to 200 pounds per square inch, and alteration of patterns for positioning or "laying up" the putty.

In addition, the putty type was changed, the reuse of motor segments increased, and a new government contractor (Lockheed) took over management of booster assembly at Kennedy Space Center. The commission stated that one of these developments or a combination of them were probably the cause of the higher anomaly rate.

A program in safety, reliability, and quality assurance should have tracked and discovered the reason for increasing erosion and blow-by, the commission said, adding: "Even the most cursory examination of the

---

[9] As related earlier, Marshall and Thiokol engineers knew that doubling leak check pressure increased incidence of blowholes through the putty, resulting in more instances of erosion and blow-by, but they feared lower pressure checks would allow putty to mask failure of an O-ring to seal once mating was completed. It was Hobson's choice.

failure rate should have indicated that a serious and potentially disastrous situation was developing on all SRB joints. Not recognizing and reporting this trend can only be described in NASA terms as a 'quality escape,' a failure of the program to preclude an avoidable problem. If the program had functioned properly, the Challenger accident might have been avoided."

Misinformed, Misrepresented, and Misled

Another failure of the agency's safety responsibility was the misrepresentation of the redundancy of the field joint sealing system, the O-rings. Although it had been categorized as criticality 1 in 1982, the Problem Assessment System that was operated by Rockwell at Marshall still listed it as criticality 1-R (redundant) on March 7, 1986, more than five weeks after the accident.

"Such misrepresentation of criticality must also be categorized as a failure of the safety, reliability and quality assurance program. As a result, informed decision making by key managers was impossible," said the report.

In the commission's judgment, top management at NASA was influenced by misinformation. The report cited as an instance an interview with the former NASA associate administrator for space flight, Jesse Moore, by a commission representative at NASA headquarters April 8, 1986. Discussing 51-B seal erosion, Moore was quoted as saying:

they reported the two erosions on the primary and some 10 or 12 per cent erosion on the secondary O-ring on the flight in April and the corrective action, I guess, that had been put in place was to increase the test pressure, I think, from 50 psi [pounds per square inch] to 200 or 100 psi—I guess . . . 200 psi is the number—and they felt that they had run a bunch of laboratory tests and analyses that showed that by increasing the pressure up to 200 psi this would minimize or eliminate erosion. And that there would be a fairly good degree of safety factor margin on erosion as a result of increasing this pressure and ensuring that the secondary O-ring had seated. And so we left that FRR [flight readiness review] with that particular action closed by the project.

The commission commented that not only was Moore "misinformed" about the effectiveness and the potential hazards of the procedure, but he was also misinformed about the issue of joint redundancy. The report stated:

Apparently no one told, or reminded, Mr. Moore that while the solid rocket booster nozzle joint was criticality 1R, the field joint was criticality 1. No one told him about blow holes in the putty, probably resulting from increased stabilization pressure and no one told him that this . . . procedure had been in use since the exact time that field joint anomalies had become dangerously frequent.

The report summarized: "Erosion was the enemy and increased [leak check] pressure was its ally. While Mr. Moore was not being intentionally deceived, he was obviously misled. The reporting system simply was not making trends, status and problems visible with sufficient accuracy and emphasis."

As an example of inadequate problem reporting and follow up, the commission cited the launch constraint imposed after erosion was found in both O-rings of the left booster nozzle joint following the launch of *Challenger* 51-B (April 29, 1985). The damage indicated that the primary O-ring had failed to seal at all.

Commission members had expressed their surprise when they learned of the launch constraint, especially when they learned also that other NASA centers and higher levels of NASA management had not been notified of it. It had no effect on the launch schedule, as mentioned earlier.

The commission explained that under agency rules a launch constraint arises from a flight safety issue of sufficient seriousness to justify a decision not to launch. The constraint following 51-B supposedly affected the next six shuttle missions. Marshall executives had ex-

plained in their testimony that the constraint was waived to allow them to fly.

The commission reported that although the NASA Problem Reporting and Corrective Action document requires project offices to inform Level 2 of launch constraints, neither that level nor Level 1 was informed.

There was other slippage in the reporting practices in the agency. In the review process preceding the launch of 51-L, two problems that had cropped up in the immediately preceding flight, 61-C *(Columbia)*, were not evaluated, the commission said. One was a serious failure of the orbiter's wheel brakes. The other was an O-ring erosion problem that had not been incorporated in the problem assessment system when 51-L was launched. The commission noted that the wheel brake problem had not been mentioned to the 51-L crew, which had planned to land at Kennedy Space Center at the end of the mission.[10]

"If the program cannot come to grips with such critical safety aspects before subsequent flights are scheduled to occur," the commission report said, "it obviously is moving too fast."

Panels of Safety

The commission recalled that the Aerospace Safety Advisory Panel was established in the aftermath of the Apollo 1 fire of January 27, 1967. It was later established by act of Congress as a senior advisory committee to NASA, with the duty of reviewing and making recommendations for safety studies and actions.

The panel began reviewing the space shuttle program in 1971 and gradually extended its scope to engineering and logistical aspects of the program. It investigated the initial proposal to use the light-weight

---

[10] The 61-C mission also was planned to land at Kennedy, but weather forced the crew to land at Edwards Air Force Base. There the extensive runway system on the Mojave Desert and calm winds made landing less hazardous than it might have been at Kennedy.

external tank and a program to replace the steel solid rocket motor case with a filament-wound case.

The commission commented: "There is no indication, however, that the details of the solid rocket booster joint design or in-flight problems were ever the subject of a panel activity. The efforts of this panel were not sufficiently specific and immediate to prevent the 51-L accident."

The commission also noted the existence of a short-lived Space Shuttle Crew Safety Panel established in 1974. It served an important function in NASA flight safety activities until it went out of existence in 1981, the commission said. "If it were still in existence, it might have identified the kinds of problems now associated with the 51-L mission," the report said.

The commission report called for a new safety organization in NASA. While the Aerospace Safety Advisory Panel had contributed to the safety of NASA operations, it could not be expected to uncover "all the potential problems nor can it be charged with failure when accidents occur that in hindsight were clearly probable."

"The commission believes therefore that a top-to-bottom emphasis on safety can best be achieved by a combination of a strong central authority and a working level panel devoted to the operational aspects of shuttle flight safety."

Specifically, the commission found that:

Reductions in the safety, reliability, and quality assurance work force at Marshall and NASA headquarters have seriously limited capability in those vital functions.

Organizational structures at Kennedy and Marshall have placed safety, reliability, and quality assurance offices under the supervision of the very organizations and activities whose efforts they are to check.

Problem-reporting requirements are not concise and fail to get critical information to the proper levels of management.

Little or no trend analysis was performed on O-ring erosion and blow-by problems.

As the flight rate increased, the Marshall safety, reliability, and quality assurance work force was decreasing, which adversely affected mission safety.

Five weeks after the 51-L accident, the criticality of the solid rocket motor field joint was still not properly documented in the problem reporting system at Marshall.

Pressure

The commission made a study of shuttle operating conditions, on the lookout for factors that put pressure on the flight system. It was the first such study by an independent investigative body since the creation of NASA in 1958. The study reported that a number of pressures were affecting shuttle operations as a result of an accelerating flight rate, hardware problems, acquiescence to late payload changes at the request of customers, and crew changes to accommodate members of Congress.

A hardware problem in a tracking and data relay satellite in the cargo bay of the *Discovery* 51-E mission resulted in canceling the mission in April 1985 and combining part of its payload with that scheduled for 51-D, launched April 12, 1985. This required deleting from 51-D the retrieval of the Long Duration Exposure Facility (LDEF), which had been deployed a year earlier with 57 experiments. The LDEF was added to STS 60-I, which was grounded after the 51-L accident. A new mission, designated as 61-M, was scheduled in July 1986 to launch a tracking and data relay satellite, replacing the one lost on 51-L. Schedule adjustments then had to be projected.

This domino effect happened repeatedly in the shuttle manifest schedule, the commission noted. Satellite customers requested changes in scheduled launch dates because of development problems, financial difficulties, or changing market conditions, the report said. As Kennedy processing people complained, this was no way to run a trucking business.

"NASA generally accedes to these requests and has never imposed penalties available," the commission commented. It cited another musical-chairs exercise affecting four flights, arising from a request to postpone the launch of a Westar satellite from 61-C.

NASA obligingly transferred the satellite to mission 61-E, scheduled for the spring of 1986. Although 61-E was grounded with the rest of the post-51-L mission schedule, the transfer required a bookkeeping entry removing from the 61-E a bridge assembly experiment. The opening on 61-C was then filled by a Hughes 376 communications satellite originally scheduled for 51-L, which then acquired the Spartan-Halley observatory scheduled for 61-D (now grounded).

The commission cited as external factors creating pressure on the system the late additions of Senator Jake Garn (R-Utah) to the crew of *Discovery* 51-D (April 12, 1985) and of Representative Bill Nelson (D-Florida) to the crew of 61-C.[11] The commission noted that a payload specialist (Nelson) was added to 61-C only two months before scheduled launch. Because there were already seven crew members assigned to the flight, one had to be removed. He was Gregory B. Jarvis, a payload specialist employed by the Hughes Aircraft Company, who was to conduct an experiment pertaining to the behavior of liquid fuel in zero or low gravity. Jarvis was moved from 61-C to 51-L just three months before 51-L was scheduled for launch, the commission reported.

The commission examined human factors affecting the efficiency and safety of shuttle flight operations. It observed that at the same time NASA was increasing the flight rate, a number of factors had the effect of reducing the number of skilled personnel to deal with it. The factors were early retirement, freezes on hiring,

transfers of skilled people to other programs, especially the space station, and the transition to a single contractor for operations support (consolidating the work previously done by several contractors).

The result of processing more flights with fewer experienced workers was overtime for a significant percentage of the 5,000 contract employees in shuttle processing. During November, December, and January (1985–86), the commission reported, monthly overtime percentages rose to 27.7 percent of the labor force in some periods and to 20–26 percent in others. The Kennedy director of space shuttle operations, Robert Sieck, told the commission that 20 percent overtime was equivalent to a 48-hour week. There was no system for monitoring overtime from a safety perspective, according to Sieck and David Owen, deputy program manager for the Lockheed Space Operations Company, the processing contractor.

The commission described one "potentially catastrophic human error" as a probable consequence of a stretched-out work week. It occurred 4 minutes 55 seconds before the scheduled liftoff of *Columbia* 61-C on January 6, 1986. Fortunately, the launch was scrubbed 31 seconds before liftoff.

According to Lockheed, approximately 18,000 pounds of liquid oxygen were inadvertently drained from the external tank because of an operator error. The liquid oxygen flow dropped the main engine inlet temperature below an acceptable limit, causing a hold in the countdown at 31 seconds and a scrub.

"Had the mission not been scrubbed," reported the contractor, "the ability of the orbiter to reach the defined orbit may have been significantly impacted."

An investigation showed that console operators in the Launch Control Center had misinterpreted system error messages resulting from a failed microswitch on a replenishment valve. "Instead of manually overriding an automatic sequencer and closing the next valve in sequence, they pressed 'continue' and caused the vent and drain valves to open prematurely."

The report cited operator fatigue as one of the

---

[11] Senator Garn, who had more hours in jet aircraft than most astronauts, was a member of the Senate Appropriations Committee and chairman of the subcommittee on HUD and Independent Agencies. Congressman Nelson represented Florida's 11th district, where the Kennedy Space Center is located, and was chairman of the House subcommittee on space science and applications.

factors contributing to the incident. The operators had been on duty at the console for 11 hours during the third day of working 12-hour night shifts (8 P.M. to 8 A.M.), the report stated.

The 61-C mission was finally launched January 12, 1986.

The Feynman Commentary

In an appendix to the commission report, Dr. Feynman wrote a critique of the shuttle program in which he raised questions about the reliability not only of the solid rocket boosters but also of the main engines and the general-purpose computers that function as the brain of the avionics or flight control system.

"It appears," he wrote, "that there are enormous differences of opinion as to the probability of a failure with loss of vehicle and of human life. The estimates range from 1 in 100 to 1 in 100,000. The higher figure [1 in 100] comes from working engineers and the very low figures from management. What are the causes and consequences of this lack of agreement?"

A loss of 1 in 100,000 would imply, he said, that the shuttle could fly every day for 300 years with only one expected loss. "What is the cause of management's fantastic faith in the machinery?"

Analyzing that state of mind, the physicist noted that certification criteria used in flight readiness reviews "often develop a gradually decreasing strictness." It is argued that the same risk was flown before without failure; this is often accepted as an argument for the safety of accepting it again.

"Because of this, obvious weaknesses are accepted again and again, sometimes without a sufficiently serious attempt to remedy them or to delay a flight because of their continued presence."

Dr. Feynman cited reports by Louis J. Ullian, the range safety officer, on the reliability of solid fuel rockets. These showed a failure rate of 1 in 25, the physicist said. But because that estimate included rockets in early stages of development, a more reasonable figure for

mature rockets might be 1 in 50. With special care, a figure of below 1 in 100 might be achieved, he said, but 1 in 1,000 is probably not attainable with today's technology. Shuttle failure rates would be double these estimates because there are two solid rockets on the shuttle.

NASA officials argue that the figure is much lower, that these figures are for unmanned rockets. But since the shuttle is a manned vehicle, the probability of mission success is "very close to 1.0."

"It is not very clear what this phrase means," Dr. Feynman said. "Does it mean it is close to 1 or that it ought to be close to 1?"

If real probability is not small, flights would show troubles, near failures, and possibly actual failures with a reasonable number of trials, he said. Standard statistical methods could give a reasonable estimate. "In fact, previous NASA experience had shown on occasion just such difficulties, near accidents and accidents, all giving warning that the probability of flight failure was not so very small. The inconsistency of the argument not to determine reliability through historical experience, as the range safety officer did, is that NASA also appeals to history [by saying], 'Historically, this high degree of mission success . . .'"

Dr. Feynman asked: "if we are to replace standard numerical probability usage with engineering judgment, why do we find an enormous disparity between the management estimate and the judgment of the engineers?"

He answered: "It would appear that for whatever purpose, be it for internal or external consumption, the management of NASA exaggerates the reliability of its product to the point of fantasy."

The Russian Roulette Syndrome

Dr. Feynman cited the *Challenger* flight as an example of "the phenomenon of accepting for flight seals that had shown erosion and blow-by in previous flights."

man

Dr. Richard P. Feynman, Nobel laureate physicist of the California Institute of Technology, criticized NASA's acceptance of seal erosion in solid rocket motors as analogous to Russian roulette. (NASA)

But erosion and blow-by are not what the design expected. They are warnings that something is wrong. The equipment is not operating as expected and therefore there is a danger that it can operate with even wider deviations in this unexpected and not thoroughly understood way.

The fact that this danger did not lead to catastrophe before is no guarantee that it will not the next time

unless it is completely understood. When playing Russian roulette, the fact that the first shot got off safely is little comfort for the next.

The origin and consequences of the erosion and blow-by were not understood. They did not occur equally on all flights and all joints; sometimes more, sometimes less. . . . In spite of these variations from case to case, officials behaved as if they understood it giving apparently logical arguments to each other, often depending on the "success" of previous flights.

For example, in determining if flight 51-L was safe to fly in the face of ring erosion on flight 51-C, it was noted that the erosion depth was only one-third of the radius. It had been noted in an experiment that cutting the ring as deep as one radius was necessary before the ring failed . . . it was [therefore] asserted there was a safety factor of three. This is a strange use of the engineer's term, safety factor.

Dr. Feynman explained by analogy: If a bridge is built to withstand a certain load, it may be designed to stand up under three times that load. This safety factor is to allow for uncertain excesses of load.

If unexpected load comes on the bridge and a crack appears in a beam—this is a failure of design. There was no safety factor at all, even though the bridge did not collapse because the crack went through only one-third of the beam. The solid rocket motor O-rings were not designed to erode. Erosion was a clue that something was wrong. It was not something "from which safety could be inferred."

Dr. Feynman then commented on the shuttle's main propulsion system, the hydrogen-oxygen engines in the orbiter. He asked rhetorically: "Were the organization weaknesses that contributed to the accident confined to the solid rocket booster sector or were they a more general characteristic of NASA?"

Referring to the main engines, he said that the usual way such engines are designed may be called the component system or bottom-up design. The properties and limitations of materials are determined. Components are designed and tested individually. As deficiencies or errors are noticed, they are corrected and corrections verified with further testing. There is a good chance that modifications to get around final difficulties are not hard to make, for the most serious problems already have been discovered and dealt with.

"But the space shuttle main engine was developed differently—top down, we might say. The engine was designed and put together all at once with little prelim-

inary study of materials and components. Then, when troubles were found in bearings, turbine blades, coolant pipes, etc., it is more expensive and difficult to discover the causes and make changes," he said.

Under these conditions, Dr. Feynman pointed out, a simple fix during top-down development may require redesign of the engine.[12]

As expected, he said, many different kinds of flaws and difficulties have turned up. Because it was built in the top-down manner, they are difficult to find and fix. The design lifetime of firings equivalent to 55 missions (27,000 seconds of operation in flight or on the test stand) has not been obtained.

The engine, he said, now requires very frequent maintenance and replacement of important parts—turbopumps, bearings, sheet metal housing, etc. The high-pressure fuel turbopump has been replaced every three or four missions, and the high-pressure oxygen turbopump every five or six. This is at most 10 percent of the original specification. In 250,000 seconds of operation, the engines have failed seriously perhaps 16 times,[13] he said.

The probability of mission failure due to main engine failure is calculated as 1 in 10,000 by Rocketdyne (the manufacturer), 1 in 300 by the Marshall Space Flight Center, and 1 in 100,000 by NASA managers, he said. An independent NASA engineering consultant believed that 1 or 2 in 100 was a reasonable estimate.

"It is evident that the flight readiness reviews and certification rules show a deterioration for some of the space shuttle main engine problems that is closely analogous to the deterioration seen in the rules for the solid rocket booster," he said.

---

[12] The development difficulties Dr. Feynman described did occur during main engine development in the late 1970s and were a major factor in delaying the shuttle's first orbital test flight more than two years. See R. S. Lewis, *The Voyages of Columbia* (New York: Columbia University Press, 1984).

[13] The entire abort-emergency landing system for the orbiter is based on the probability of main engine failure only, but no missions were aborted for this reason. Failures resulting in explosions occurred during testing.

Dr. Feynman next considered shuttle avionics, citing the general computer system with 250,000 lines of code as highly elaborate. He explained that it is responsible for the automatic control of the entire ascent to orbit and for the descent well into the atmosphere below Mach 1.

There is not enough room in the memory of the main-line computers for all the programs, Dr. Feynman said, of ascent, descent, and payload operations in flight; consequently the astronauts load the memory about four times from tapes. He said that no change has been made in the hardware system in 15 years. The actual hardware is obsolete. Memories are of the old ferrite-core type. It is becoming more difficult to find manufacturers to supply such old-fashioned computers reliably.

"Modern computers," he said, "are more reliable, can run faster, simplifying circuits and allowing more to be done and would not require so much loading of memory since their memories are much larger."

"Such a Dangerous Machine . . ."

Dr. Feynman concluded that in order to keep up with the launch schedule, criteria are altered so that flights may still be certified in time. "They therefore fly in a relatively unsafe condition with a chance of failure of the order of a percent. It is difficult to be more accurate."

He added: "Official management, on the other hand, claims to believe the probability of failure is a thousand times less. One reason for this may be an attempt to assure the government of NASA's perfection and success in order to ensure a supply of funds. The other may be they sincerely believe it to be true, demonstrating an almost incredible lack of communication between themselves and the working engineers.

"In any event, this has had very unfortunate consequences, the most serious of which is to encourage ordinary citizens to fly in such a dangerous machine as if it had attained the safety of an ordinary airliner . . .

"Let us make recommendation that ensure that if NASA officials deal in a world of reality in understanding technological weaknesses and imperfections well enough to be actively trying to eliminate them, only realistic flight schedules should be proposed, schedules with a reasonable chance of being met.

"NASA owes it to the citizens from whom it asks support to be frank and honest and informative so that those citizens can make the wisest decision for the use of their limited resources.

"For a successful technology, reality must take precedence over public relations, for nature cannot be fooled."

10. Transfiguration

The conclusion of the Presidential Commission's investigation of the *Challenger* accident was a series of nine recommendations that were adopted unanimously "to help assure the return to safe flight." The commission urged that the administrator of NASA submit a report by June 6, 1987 to the President on the progress that NASA had made in effecting the recommendations.

Recommendation 1 called for the redesign of the faulty solid rocket motor joints—either a new design eliminating the joint or a redesign of the current joint and seal. It stated that no options should be precluded because of schedule, cost, or reliance on existing hardware. It specified that:

Joints should be fully understood, tested, and verified. The integrity of the seals should not be less than that of the case walls. The integrity of the joints should be insensitive to dimensional tolerances, transportation and handling, test procedures and inspections, environmental effects, recovery and reuse, and flight and water impact loads.

The certification of the new design should include tests that duplicate the actual launch configuration as closely as possible and cover the full range of operating conditions, including temperatures.

To make sure that these recommendations are carried out, the NASA administrator should ask the National Research Council to form an independent oversight committee to oversee the solid rocket motor improvements. The oversight committee should report to the administrator on the adequacy of the design and make appropriate recommendations.

Recommendation 2 dealt with the management of the shuttle program. It said that a new definition of the program manager's responsibility is essential, noting that the project managers for the various elements of the shuttle "felt more accountable to their center management than to the shuttle program organization." As a result, vital information frequently bypasses the national shuttle program manager. Program funding and all program work at the centers should be placed under the program manager's authority.

NASA should encourage the transition of astronauts into agency management and elevate the function of the flight crew operations director in the agency's organization.

NASA should establish a shuttle safety advisory panel that would report to the shuttle program manager. Representation on the panel would include agency's safety organization, mission operations, and the astronaut office.

Recommendation 3 called for a review of the critical items list, ranging from 1 and 1-R to 2 and 2-R. The review should identify items that must be improved before flight to ensure safety. It was proposed that an audit panel, appointed by the National Research Council, should be installed to verify the adequacy of the criticality and hazard review. The panel would report directly to the administrator.

Recommendation 4 called for the establishment of the office of safety, reliability and quality assurance to be headed by an associate administrator, reporting to the administrator. The office should be sufficiently staffed

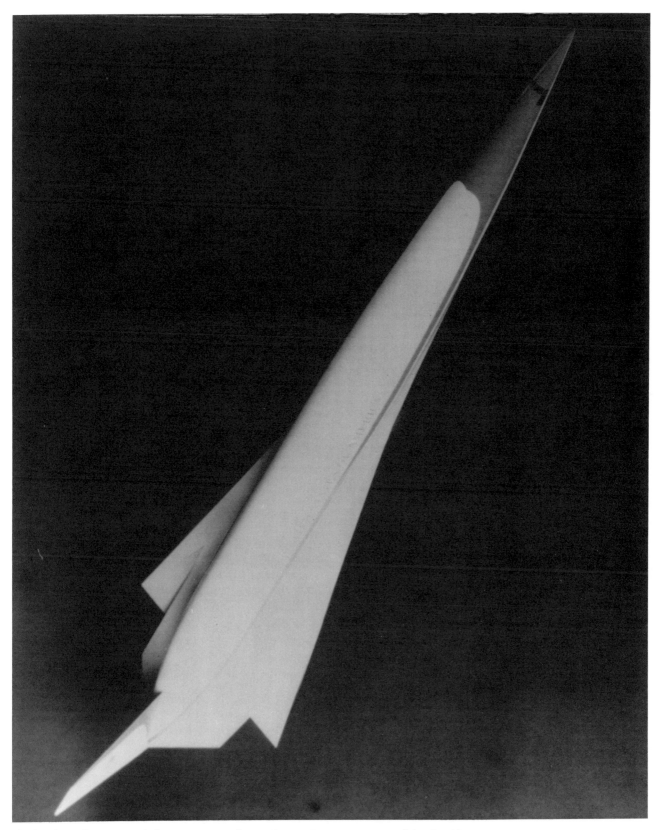

Idealized configuration of the aerospace plane "Orient Express" suggested by President Reagan as a possible development beyond the shuttle. It would take off horizontally, like an airplane, and accelerate into orbit. After reentry, it would land on a conventional runway. It would also be able to cruise hypersonically in the atmosphere on intercontinental passenger flights. The British are at work on a similar concept called HOTOL—horizontal take-off and landing aerospacecraft. (NASA)

to ensure effective oversight of safety, quality assurance, and reliability programs. It should be independent of other agency functional and program responsibilities.

Recommendation 5 demanded an end to what the commission called "management isolation" at the Marshall Space Flight Center. It said that the commission found that Marshall project managers failed to provide full and timely information bearing on the safety of flight 51-L to other vital elements of shuttle program management. The recommendation said that NASA should take energetic steps to eliminate this tendency toward management isolation at the center "whether by changes of personnel, organization, indoctrination—or all three."

The recommendation called for a policy governing the imposition of shuttle launch constraints, for the recording of flight readiness reviews and mission management team meetings, and for the attendance of the flight crew commander or his designated representative at the flight readiness review. The commander or representative should be a party to accepting the vehicle for flight and certify that the crew is properly prepared for flight.

Recommendation 6 urged NASA to improve landing safety. Tire, brake, and nosewheel steering systems must be improved, it said, adding: "These systems do not have sufficient safety margin, particularly at abort landing sites."

Specific conditions under which planned landings at Kennedy would be acceptable should be determined. Criteria must be established for tires, brakes, and nosewheel steering, and until these criteria are met, landing at Kennedy should not be planned. During unpredictable weather periods at Kennedy, landings should be planned at Edwards Air Force Base, California.

Recommendation 7 called upon NASA to "make all efforts to provide a crew escape system for use during controlled gliding flight." The commission stopped short of urging a launch escape system. It said: "The shuttle program management considered first-stage abort options and crew escape options several times during the history of the program, but because of limited utility, technical infeasibility or program cost or schedule, no systems were implemented."

The commission recommended that NASA seek to increase the range of flight conditions under which an emergency runway landing can be made when two or three main engines fail early in the ascent. It did not indicate how this could be done.

Recommendation 8 stated that NASA must establish a flight rate "consistent with its resources" and establish a firm payload policy with controls on manifest changes to reduce pressure on schedules and crew training. It urged NASA to avoid reliance on a single-launch capability in the future, because "The nation's reliance on the shuttle as its principal space launch capability created a relentless pressure on NASA to increase the flight rate." The recommendation implied that NASA should shift commercial payloads to expendable rockets to relieve pressure on the shuttle.

Recommendation 9 called for the establishment of a system of analyzing and reporting performance trends for criticality 1 items. Maintenance procedures for these nonredundant items should be specified in the critical items list, especially for the shuttle main engines. Also recommended for the orbiter was a comprehensive maintenance inspection plan, periodic structural inspections that could not be waived, and cessation of cannibalizing parts from one orbiter to repair another. The spare parts inventory should be restored and maintained.

Skirting Recommendations

The recommendations, addressed by protocol to the President, were directed to the reappointed administrator of NASA, James C. Fletcher. In a statement on June 9, 1986, he declared that "we have been pressing on, despite the pain, seeking answers to difficult questions; beginning carefully to make changes where they are needed. We have been at work. Yet, like all Americans, we have awaited the Rogers Commission report, hoping to learn from it as well."

Fletcher noted that the report of a presidentially appointed, independent body carried special status "and the compelling obligation to study its conclusions with great care. We are prepared to do that with an open mind and without reservations."

But the Rogers commission report imposed a considerably stronger response than an obligation to study its conclusions "with great care" and "with an open mind." In a "Dear Jim," letter to Fletcher dated June 13, President Reagan told the administrator to implement the commission's recommendations "as soon as possible." The letter, in effect an order, stated: "The procedural and organizational changes suggested in the report will be essential to resuming effective and efficient space transportation system operations, and will be crucial in restoring U.S. space launch activities to full operational status."

Fletcher submitted a plan to the White House on July 14, 1986 to implement all the commission's recommendations and report on the results by June 1987.

The stage was set and arrangements were in place for NASA's transfiguration. Fletcher declared: "Where management is weak, we will strengthen it. Where engineering or design or process need improving, we will improve them. Where our internal communications are poor, we will see that they get better."[1]

Before long, however, it became apparent that NASA was going to sidestep two key recommendations for the redesign and testing of the solid rocket motors.

One of them was the commission's warning against precluding design options for reasons of schedule, cost, or reliance on existing hardware. By summer, NASA adopted fixes for the booster joints precisely for those

reasons. The redesign would use motor cases already in inventory, saving millions of dollars, and would allow the resumption of flight at an earlier date than options that require replacement of existing hardware or redesign of a non-segmented booster.

The other recommendation asked that consideration be given to testing redesigned booster joints in their full configuration, that is, in a vertical position. The boosters had been test-fired previously in a horizontal position. During commission hearings, critics of horizontal testing asserted that it did not put the same weight on the joints that a vertical test would and thus did not impose the conditions of actual launch.

As the commission had urged, NASA did consider vertical testing—but rejected it. Vertical test stand facilities would have had to be built at a cost estimated at $30 million. The construction would delay the resumption of flight an extra year. Besides, engineering opinion was divided on the efficacy of vertical vs. horizontal rocket testing. NASA decided to stay horizontal.

The Old Hands Return

Among current and former astronauts, the commission report was well received. Former U.S. Senator Harrison H. Schmitt (R-N.M.), an ex-astronaut who flew to the Moon in 1972 on Apollo 17, last of the lunar landing missions, praised commission report. In many respects, it echoed the views he had expressed on the shortcomings in shuttle development, as a member of the Senate subcommittee on science, technology, and space (1977–83). Schmitt, as ranking minority member, and the subcommittee chairman, former Senator Adlai E. Stevenson (D-Ill.), had been critical of NASA cost-shaving development policy and all-up testing of the orbiter main engine, which set the shuttle flight program back two years.

Schmitt, a geologist, was the only scientist who set foot on the Moon during the six landings, and his views

---

[1] The administrator's implementation plan was a 50-page document prepared by Admiral Truly's staff entitled: "Actions to Implement the Recommendations."

on the origin and evolution of the Moon were widely respected. Now he came forward to say that the commission's report "is a tribute to the diligence and insights of the commission's members and a fitting legacy for action in behalf of the Challenger crew." [2]

On one point, however he took exception: the commission's recommendation to set up the office of associate administrator for safety, reliability, and quality control in NASA headquarters. "The focus of this responsibility on a single high level manager may make it bureaucratically acceptable to avoid such responsibility at other levels of management," he said. He cited the views of Hans Mark, a former NASA deputy administrator, who held that safety, reliability, and quality control were so intrinsic to the success of complex ventures that all levels of NASA must be held responsible for them.

Although the commission's dictates were not to be followed in all respects, their impact deflated and impelled reorganization in a NASA bureaucracy that had been credited with some of the greatest achievements in the history of technology. Essentially, the top levels of the bureaucracy were removed by resignation, reassignment, or retirement. Most vacancies were filled by "old hands" who understood NASA administration and technology. The top "old hand" was Fletcher. His return as administrator, the job he had under Nixon and Ford, appeared significant in the light of the shuttle's rising military role.

In addition to his career in the aerospace industry, Fletcher had served extensively as a defense consultant. He was an assistant Secretary of the Air Force as well as consultant to the Secretary of Defense, a member of the Air Force Science Advisory Board and of the President's Science Advisory Committee, a member of the Strategic Weapons Panel, chairman of the Naval Warfare Panel and member of the Defense Science Board. More recently, he headed the Defensive Technologies Study Team (the Fletcher commission), which confirmed a research agenda for the Strategic Defense Initiative.

William R. Graham, acting NASA administrator since the departure of James M. Beggs, was appointed by the President as the White House science adviser, replacing George Keyworth, who had resigned in 1985 to become a consultant. Graham's initial post as deputy administrator under Beggs was filled by Dale D. Myers, another "old hand." Myers, a former executive at North American Rockwell in the Apollo era, had been NASA associate administrator for manned space flight from 1970 to 1974 and had played a part in shuttle development.

Another "old hand," Major General Samuel C. Phillips, Air Force, retired, was asked to come back to headquarters as a management consultant and draft an organization plan. As Apollo general manager in the 1960's he warned repeatedly of program slippage and cost overruns preceding the 1967 Apollo fire.

By late summer, astronauts were moving into executive posts at headquarters. In addition to Truly, now associate administrator for space flight, Navy Captain Rick Hauck was named associate administrator for external relations; Air Force Colonel Frederick D. Gregory was appointed chief of operational safety in the new Office of Safety, Reliability, and Quality Assurance; and Dr. Sally K. Ride, the then current astronaut on the Rogers commission, was named as special assistant to the administrator. To the NASA administration these new officials brought a flight crew's insight into the capabilities and problems of the shuttle orbiter. They instilled an operations viewpoint into a command hierarchy that had been virtually the exclusive province of engineers. [3] More changes were soon to follow.

Jesse Moore, who had assumed his pre-accident appointment as Johnson Space Center director when

---

[2] Schmitt's comment on the commission report was written exclusively for the *Orlando Sentinel* and published June 15, 1986.

[3] Navy Captain Robert L. Crippen was appointed to head a team that would monitor shuttle program management. Two months later, he was named deputy director of shuttle operations. NASA announced in May 1987 that Dr. Ride planned to resign from the agency as of August 17 to become a research fellow at Stanford University.

Sally Ride

Astronaut Sally Ride, a Ph.D. in physics and the first American woman to fly in space, as a member of the Rogers commission. She agreed to serve in NASA headquarters as special assistant to Administrator James C. Fletcher. (NASA)

Truly came aboard at headquarters, decided to leave after several months. He asked for reassignment to headquarters as assistant to the NASA general manager, Philip Culbertson, but first he requested sabbatical leave. He was quoted as saying that he wanted time to recover from the emotional stress of the *Challenger* accident and also the strain of dealing with conflicts over space station development. Moore resigned from NASA on

February 8, 1987 to become director for program development at Bell Aerospace Systems, Boulder Colorado.

Meanwhile, George A. Rodney, 65, ex–test pilot and engineering executive at Martin Marietta, Orlando, was appointed NASA associate administrator for Safety, Reliability, and Quality Assurance, the add-on headquarters function recommended by the Rogers Commission. He had supervised quality control for Martin Marietta at the Michoud, Louisiana plant where the external tank was manufactured.

At the Kennedy Space Center, Richard G. Smith retired as director in July 1986 after seven years in that job. He was a 35-year veteran of space programs, having begun a career in rocket research and development at the Redstone Arsenal upon being graduated from Auburn University. Redstone was the site of rocket development by the Army Ballistic Missile Agency whose engineers formed the nucleus of the Marshall Space Flight Center. Smith was transferred to NASA in 1960. In January 1974, he became deputy director of Marshall and in August 1978, deputy associate administrator for space transportation systems at NASA headquarters.

One of Smith's notable accomplishments was his management of the Skylab Task Force that maneuvered the falling Skylab space station so that it plunged into the Pacific Ocean off the coast of Australia on July 11, 1979 without damaging a populated area. Abandoned in early 1974, its experimental mission completed, after being occupied by three crews for a total of 171 days, the huge station, 86 feet long and with a mass of 170,000 pounds, began to fall out of orbit in 1977, years before its decent was predicted. It had been proposed that the shuttle be used to boost Skylab into a higher, more stable orbit, but the shuttle could not be made ready in time.

Smith became president and chief executive officer of the General Space Corporation, Pittsburgh, Pennsylvania. He was succeeded by Lieutenant General Forrest S. McCartney, 55, who had been serving as commander of the Air Force Space Division. McCartney had a varied career as a nuclear engineer and weapons specialist and had been director of range engineering at the Air Force Eastern Test Range (Cape Canaveral) from 1971 to 1974.

At the Marshall Space Flight Center, William R. Lucas retired as center director on July 3 and was succeeded in August by James R. Thompson, 50, the former associate director of engineering at the center,

who had returned to NASA from Princeton University to assist Truly as vice chairman of the Design and Data Analysis Task Force. Now he undertook the direction of Marshall in its most critical period.

In May, Lawrence B. Mulloy, Marshall's booster project manager who had opposed the Morton Thiokol engineers' no-launch recommendation, was reassigned as deputy director of science and engineering. That post had been vacated by George B. Hardy, who retired May 2. Hardy had supported Mulloy's contention that it was safe to launch *Challenger*.

At Morton Thiokol, a similar reshuffling of jobs took place. Gerald Mason, senior vice president of the company's Wasatch Operations, retired. Two company vice presidents, Joseph Kilminster and Calvin G. Wiggins, were reassigned. All had participated in management's launch decision.

Gerald W. Smith, a veteran Marshall engineer who had been acting propulsion chief at headquarters, succeeded Mulloy as booster manager at Marshall in Huntsville. Shortly thereafter, Mulloy retired. Stanley Reinartz, who as shuttle projects office manager had been Mulloy's boss, was reassigned as manager of the special projects office. William R. Marshall became shuttle projects manager.

Meanwhile, the priority effort of redesigning the solid rocket motor joints and seals was placed under the management of John W. Thomas, spacelab program office manager at Marshall, by Admiral Truly. James E. Kingsbury, who initially had headed this program, returned to his former position as director of science and engineering. On November 30, he retired.

When the parade of musical chairs ended at Marshall in May, management authority had been realigned to comply with the recommendations of the Rogers commission. Henceforth, it would function under a tighter rein from headquarters than the Marshall center had experienced before. In announcing May 12 that Thomas would assume management of the solid rocket motor redesign team, Truly said that an independent group of senior experts would be formed to oversee the motor redesign. It would report to the administrator of NASA. A panel on booster joint redesign was appointed by the National Research Council's Commission of Engineering and Technical Systems as an oversight committee.

At a news briefing on August 12, John Thomas reported that the preliminary booster redesign review would be completed by September 30, and the most important change was the redesign of the field joint. Two other areas would be improved—the case-to-nozzle joint and the nozzle, he said.

In summary, the redesign of the field joint added a third O-ring, eliminated the troublesome putty, which would be replaced by bonded insulation, and added a capture device. The device would prevent or at least reduce the opening of the joint (so-called joint rotation) as the booster inflated under motor gas pressure during the ignition transient. The third O-ring would be added to seal the joint at the capture device. The device would inhibit joint rotation by clamping the tang and the inner leg of the clevis together.

The current Viton or fluorocarbon O-rings would be replaced by rings of the same size (0.28 inches thick) made of fluorosilicone or nitrile rubber. Their specifications would call for resilience at 31° F. However, 2-kilowatt heating strips, very thin and about an inch wide, would be affixed to the joints to keep the O-rings at a "comfortable" temperature of 75° irrespective of the chill outside. Preliminary tests indicated that the heating strips were capable of raising booster O-ring temperature as much as 60 degrees, Thomas said. In the redesign, the tolerances were considerably improved. Gap openings that the O-rings were designed to seal were to be reduced from 30 to 6 thousandths of an inch, Thomas said. In addition to redesigning the field joints, the nozzle-to-case joint, with a history of leakage, would be strengthened by adding a bolt to fix the O-rings in place and by adding seals under 100 boltheads to stop leaks. Nozzle joint seals also were being improved with new O-rings.

Having outlined these "fixes," Thomas explained the rationale of the redesign program. "We were charged in the redesign effort with two objectives," he said. "The primary objective was to design a safe joint and the secondary objective was to utilize existing hardware if we could design a safe joint in that pursuit. I believe we have done that."

Thomas' redesign team appeared to be unanimous in establishing a rationale favoring horizontal booster firing tests instead of vertical tests. The reported consensus held that certain bending and buffeting loads on the booster could be simulated more accurately in the horizontal test stand.

Whatever technical advantages vertical testing might offer appeared to be outweighed, in NASA's view, by the money and time it would take to build the vertical test stand, presumably at Marshall.

Time was of the essence. Thomas estimated that vertical testing would add 10 to 20 months to the redesign program beyond the February 1988 date for resumption of flight.[4]

"Right now, we're talking in the area of $300 million through the certification process to get ready to fly again," Thomas said.

This price took into account that the 360 motor segments in inventory at Morton Thiokol would be usable in the new design. These constituted the existing hardware that had been incorporated in the redesign under the constraints of time and money.

"We can use all of the motor case segments we have in inventory," Thomas said. "And we only make a few additional segments that are at field joint locations. We're not discarding any existing segments."

Beyond the redesign plan Thomas described, the National Research Council's oversight panel and asked four other booster propulsion companies (other than Morton Thiokol, which was participating in the NASA redesign) to offer competing designs. They were Aerojet, United Technologies, Hercules Powder Company, and

---

[4] NASA's shuttle mission program for 1988 was based on resumption of flight on February 18, 1988.

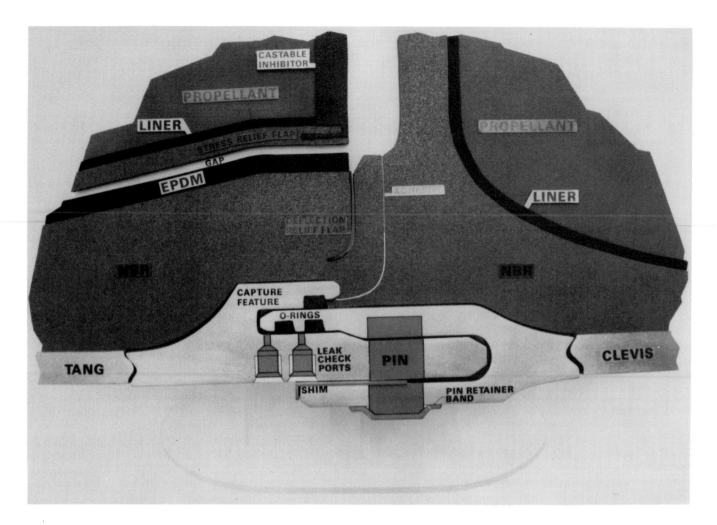

Morton Thiokol, Inc. made this model of a cross-section of the redesigned field joint and planned to test it in Development Motor 8 during July 1987. The cross-section shows the tang (tongue) as it fits into the U-shaped clevis, forming the joint. At outer left, a retainer band holds one of the 177 pins that fasten tang and clevis together around their entire circumference to secure the joint as the motor segments are mated.

Actually, the tang is a circumferential extrusion or tongue on the lower end of the booster motor segment. It fits into the circumferential slot of the clevis on the upper end of the segment. To ensure a tight fit, thin strips of metal called shims are wedged between the outer leg of the clevis and the tang at each pin.

The modifications of the redesigned joint are a third O-ring, a second leak check port, and a capture device, all additions to the earlier design. In addition, putty that formed a heat barrier in the earlier joint design has been eliminated. In its place, engineers devised the equivalent of a continuous layer of insulation across the joint. An extension of the NBR (nitrile butadiene rubber) insulation on the tang side of the joint (X) meets with a mating surface on the clevis side of the insulation. The mating surface is coated with an adhesive which bonds the two together during assembly.

Additionally, a J-shaped relief flap on the tang side of the insulation allows pressurized gases to exert increased force on the bonded layer during ignition, sealing the insulation more tightly. There is a short, nonbonded area of the mating surface that allows disassembly of the motor if necessary, without damage to the insulation surfaces.

In the old design, testing the O-rings for leaks by forcing nitrogen gas through the leak check port under pressure unseated the primary O-ring. It was assumed that the pressure of motor gas at ignition would force the primary O-ring back into its groove, but that process did not always prevent erosion. A second leak check port was added to enable technicians to reseat the primary O-ring by gas pressure during testing.

Another event allowed by the old joint design was joint rotation, forcing the inner leg of the clevis to separate from the tang at the start of motor ignition when gas pressure inflates the motor segment. This effect also displaced the primary O-ring, allowing hot gas to blow by and erode the seal. The addition of the capture feature, an extension of the tang, was designed to clamp the inner leg of the clevis to the tang to prevent joint rotation. The capture feature is fitted with the clevis to provide a metal-to-metal seal.

The function of the third O-ring is to provide a thermal barrier to protect the primary O-ring in case the bonded insulation layer fails—an event Thiokol regards as unlikely.

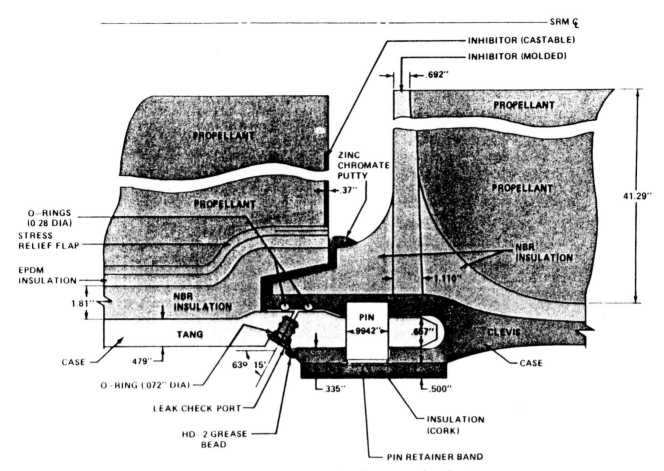

The older field joint has one leak check port, two O-rings, and used putty as a heat barrier.
(*Commission Report,* vol. 2, p. L-59)

Atlantic Research Corporation. These would be so-called "clean sheet" designs, that is, designs not using any existing hardware.

By late summer, however, NASA headquarters had virtually adopted the redesign formulated by Thomas and his team. During a visit to the Kennedy Space Center on August 18, 1986, Fletcher said that the Thomas team fix looked promising. The administrator estimated then that the cost of the *Challenger* accident would reach $630 million plus $108 million to fix the "anomaly"—the defective joint design.

The Panel Reports

In a review of the motor redesign program on October 10, 1986, the National Research Council's oversight panel concluded that "the chances for success for the current approach to case field joint redesign are sufficiently good that it should be pursued."

The panel added, however, this qualification: "The choice is the consequence of the understandable desire to use existing hardware to the greatest extent possible, including new case forgings previously ordered." It said that if the 360 case forgings had not been "a design constraint . . . we believe that more basic alternatives to the baseline design would probably be preferred once thoroughly analyzed."

The panel noted that the NASA redesign team had recommended horizontal rather than vertical firing tests and agreed that horizontal testing "can be appropriate for this situation." It added that the test program should reveal not only launch and flight loads but how these loads are transferred from the booster to the external

Primary Concerns (Cont)

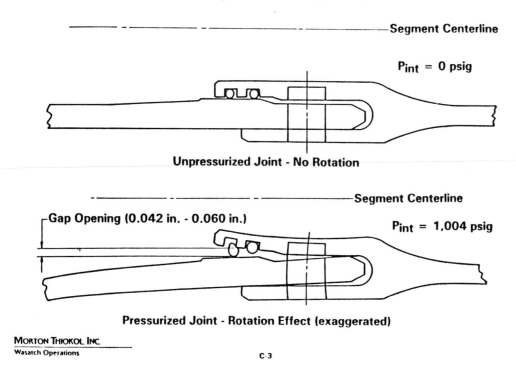

Segment Centerline

P_{int} = 0 psig

Unpressurized Joint - No Rotation

Segment Centerline

Gap Opening (0.042 in. - 0.060 in.)

P_{int} = 1,004 psig

Pressurized Joint - Rotation Effect (exaggerated)

MORTON THIOKOL, INC.
Wasatch Operations

C-3

This sketch illustrates the effect of joint rotation in displacing the primary O-ring during the ignition transient. The capture device was added to prevent this effect. (Morton Thiokol, Inc.)

tank through the fore and aft attachment struts. (The aft strut burned through on the *Challenger* flight, precipitating the main rupture of the external tank.) The redesign program should also take into account the effect on the joints of the phenomenon called "twang," in which the shuttle structure bends and then snaps back like a bow at liftoff.[5]

To assure that all such effects are observed in the test program, the panel recommended the construction of an additional horizontal test stand for full-scale motor testing. It would supplement an existing stand at the Morton Thiokol test facility at Brigham City, Utah. It would reduce the risk of a long program delay if a catastrophic accident should destroy the lone test stand.

[5] In his memorandum of December 29, 1986, John W. Young had expressed concern about the dynamic effects of wind shears during launch on the booster attach ring structure. He said that the booster test program should analyze these effects.

NASA and Morton Thiokol agreed to construct the test facility at the Thiokol plant, Brigham City, Utah, sharing the estimated $20 million cost.

Although the panel approved the capture feature, it warned that the device would increase the rigidity of the joint and cause additional bending stresses on either side. These stresses need to be understood, the panel said.

In the main, the principal features of the redesign were not new. A capture feature, a third O-ring, and elimination of the putty had been considered by a Morton Thiokol redesign group in 1985, along with other options for improving the field joints. Marshall engineers had participated in these ideas, in critiquing and analyzing them, but they never flew.

While the panel approved the redesign approach, its report pointed out alternative possibilities. The chairman, H. Guyford Stever, had served as a consultant on aerospace problems for both NASA and the Department of Defense. It was possible, he said, to redesign both case-to-case and case-to-nozzle joints without a capture feature. This could be done by modifying the simple tang and clevis joints by changing the dimensions of the parts. That would have the effect of reducing joint rotation substantially.

The concept of a joint that rotates closed or does not rotate at all might be implemented for the case joint either with bolted flanges, as suggested by NASA's Langley Research Center, or with a pinned tang-clevis joint, the report said. Either method might result in reducing or eliminating joint rotation by design rather than mechanically by the capture feature.[6]

Although the panel said it recognized the importance of returning the shuttle fleet to service without unnecessary delays, it added that ''We strongly recommend that NASA maintain a program to explore the development of original, possibly quite different designs, for the next generation of solid rocket boosters in parallel with the current redesign effort and for the contingency that the baseline design may not offer sufficiently good performance and margin of safety.''

NASA was already pursuing alternate design and development of the orbiter main engine turbopumps. In August, it announced selection of the Pratt and Whitney aircraft division, United Technologies, for contract negotiations leading to a cost plus award fee contract for the redesigned pumps. Pratt and Whitney had estimated the cost at $182 million over 5 to 6 years. The pumps would be interchangeable with those manufactured by Rocketdyne, and the redesign would attempt to prolong their service life.

Before NASA's 1986 year of disaster was out, it became evident that the shuttle program was undergoing a sea change. NASA after the accident was not the same organization as NASA before the accident. Not only was there a new team in the backfield, but the

[6] The capture device would add 600 pounds per booster, reducing the orbiter payload capacity by 100 pounds, according to *Aviation Week and Space Technology* of August 18, 1986.

style of the game and its main objectives had changed. All this happened rapidly, mainly within the 120 days of the Rogers commission investigation.

The commission's relentless probing had brought into the perspective of hindsight an impression of managerial rashness in the acceleration of a flight program beyond the system's resources of personnel and dependable hardware. The system objective of increasing the flight rate to 24 a year (the limit imposed by the external tank manufacturing capacity) by 1988 was based on the illusion that the shuttle was operational, a fiction conceded by space agency managers as early as 1984. The economic incentive of demonstrating cost-effectiveness had long since proved unrealizable, and the alternate rationale for accelerating the flight rate was to maintain world leadership in commercial and scientific space transportation.

When this frenetic effort came to a halt on January 28, 1986, a time of agonizing and guilt-ridden reappraisal set in. What was the essential purpose of the shuttle—which had become the National Space Transportation System? When it was proposed by a Presidential Space Task Group in 1969, the shuttle was presented as a reusable transport serving a space station and ultimately becoming a unit in a lunar and interplanetary transportation system. The long-range objective of such a system was to land human explorers on Mars by 1988.

But in the economically and politically troubled 1970's, that vision became blurred by changing priorities resulting from the war in Southeast Asia, racial tension, and the rising costs of federal entitlement programs. The euphoria of the lunar landings faded. Congress, the administration, and much of the news media were tending to view the manned space program as a costly distraction from more mundane concerns.

The space shuttle, having lost its priority, became something of an orphan of a moribund space policy. And when the space station was deferred indefinitely, the shuttle had nowhere to go. Its only raison d'être was to maintain an American presence in low Earth orbit where the presence of the Soviet Union was growing.

NASA then invented another rationale for a reusable manned space vehicle. It would serve as a cost-effective transport for satellites, displacing expendable rockets and thereby saving billions in launch costs. The thesis of cost-effectiveness as the economic reason for developing the shuttle was presented to Congress with evangelical fervor. It was endorsed by some economists, ridiculed by others.

Unanticipated increase in development costs, unexpected delays, and much higher than expected maintenance and processing expenses defeated the prospect of a cost-effective system. That became evident after the first four test flights. By the end of 1985, the utility of the shuttle as a cut-rate satellite carrier, vis-à-vis expendable launch vehicles, had become obviously factitious.

But other uses for the shuttle were emerging. The Department of Defense, which had supported the shuttle from the beginning and even influenced its design, had found it to be an essential asset for the national defense. It was an asset particularly in the development of the Strategic Defense Initiative as well as in the deployment of heavy military satellites.

With President's Reagan's decision to develop the permanent manned space station, the shuttle acquired a new importance as the principal construction vehicle. President Reagan's authorization of the replacement orbiter for *Challenger* reflected this function as a new priority for the shuttle fleet.

By 1986, the shuttle's original role as a means of servicing a space station and been restored, and with it, the 1969 concept of a manned lunar-interplanetary transportation system had come full circle. The shuttle was the primary transport for such a system, in which the station would serve as an operating base.

The New Flight Program

The new presidential space policy cut back the shuttle's commercial satellite-hauling business, which was to be transferred to expendable rockets. The shuttle, as of the end of 1986, was to be used principally for military, scientific, and space station construction payloads, probably for the balance of the century.

The space station was redesigned, and its development was centralized at NASA headquarters in Washington. Andrew Stofan, director of the Lewis Research Center in Cleveland, NASA's most advanced engine development facility, was tapped by Administrator Fletcher to take charge of the space station program as associate administrator at headquarters.

These changes were followed by the announcement on October 3, 1986 of the new shuttle flight program. It was scheduled to start February 18, 1988 with the launch of *Discovery*. Five flights were planned in 1988, 10 in 1989, and 11 in 1990. The flight rate would rise to 16 flights a year by 1994, at least half of them dedicated to space station construction and testing.

Under the Reagan administration's new policy, most commercial space cargo would be shifted to private-sector expendable launchers. The policy would allow NASA to fly only 19 of 44 commercial payloads under contract as of the first quarter of 1986. The 25 payloads that were canceled under the new policy would have to fly on U.S. private, French, Chinese, Soviet, possibly Japanese rockets. The European Space Agency's Ariane rocket, operated by France's Arianespace, was expected to get the largest share of the rejects. The Soviet Union created the state-controlled Cosmos enterprise to lease and launch satellites, starting with a natural resources satellite for India in 1987. China *Daily* reported in May that the Chinese Ministry of Astronautics had signed an agreement with a Texas-based joint venture, Teresat, Inc. and the China Great Wall Industry Corporation to relaunch two used American communications satellites on Long March rockets.[7] The satellites were Westar 6 and Palapa B, which had been deployed in orbit originally by *Challenger* 41-B in February 1984 but had not

---

[7] United Press International, Peking, May 13, 1986.

With the cutback in future shuttle commercial payloads ordered by President Reagan, a fleet of expendable launch vehicles is being built up. This is a version of Martin Marietta's Titan III which the corporation is building for the Air Force. With minimal modifications, the corporation said the vehicle can carry shuttle-class commercial payloads. (Martin Marietta)

reached geostationary orbit as a result of the failures of their upper-stage boosters. The two satellites had been retrieved by the crew of *Discovery* 51-A in November 1984, returned to earth, refurbished, and put on the market. China had also contracted with the Swedish Space Corporation in March 1986 to launch the Swedish Mailstar satellite into low Earth orbit.

Although increasing demands by the Department of Defense for cargo space on the shuttle threatened to curtail its commercial and scientific payloads before the *Challenger* accident, the postaccident space policy extinguished the doctrine of cost-effective launch service that NASA had promulgated in the 1970s to sell a $5.15 billion shuttle development project to congress. Yet, in spite of the fact that this blue-sky doctrine was never realized, the shuttle did succeed in making the United States the preeminent commercial space faring power.

On its 20 payload missions,[8] the shuttle had deployed 25 commercial satellites for American and foreign interests. The latter included Canada, India, Indonesia, Mexico, and Saudi Arabia. A West German spacelab had also been flown. In addition, two satellites were repaired in orbit (the Solar Maximum Mission Observatory and Leasat 3), plus the two retrieved. The shuttle thus demonstrated a capability of servicing satellites in orbit that no other space faring system could match.

On the launch ambitiously set for February 18, *Discovery* would carry a tracking and data relay satellite (TDRS), the second of three in a new space-to-ground communication system with a ground terminal at White Sands, New Mexico. The first such satellite, TDRS-1, had been deployed in low Earth orbit in April 1983 on the first flight of *Challenger*. Although its upper-stage booster had failed, the satellite had been maneuvered into geostationary orbit by auxiliary thrusters. The second TDRS had been lost aboard *Challenger* 51-L.

The third TDRS was scheduled for deployment by *Discovery* in September 1988. The full tracking and data relay satellite system would then be ready to relay data from the Hubble Space Telescope, which was to be deployed in low Earth orbit by *Atlantis* in November. The telescope was designed to peer deeper into time and space than any astronomical instrument known. It was to have been launched in the spring of 1986.

The three communications satellites would complete a new communications and tracking system that would largely replace the old global network of ground stations, inaugurated in Project Mercury a quarter of a century ago. The new system would provide continuous, 24-hour communications with the shuttle and the space station.

Two of the five 1988 flights were dedicated to the

Department of Defense on *Atlantis* in May and *Columbia* in July. Up to that time, only two dedicated defense missions had been flown, both in 1985 (51-C and 51-J). In 1989, according to the projected manifest, 4 of the 10 missions to be flown that year would be dedicated to Defense.

The projected manifest complied with the administration's new space policy; according to Fletcher, only those commercial and foreign payloads deemed to be "shuttle unique" or to have national security or foreign policy implication were considered.

From the resumption of shuttle flight to 1994, 41 percent of the cargo space would be allocated to Defense, 47 percent to NASA's needs, and 12 percent to commercial, foreign government, and U.S. government civil space requirements. The NASA allocation included scientific as well as space station payloads.

Increasing Defense use of the shuttle and a more conspicuous military presence in top NASA positions underscored the transition in the shuttle's functions. Its role as a common carrier for commercial satellites and for commercial, industrial, and scientific experiments was no longer reflected by the postaccident manifest with its increased Defense use.

Restructuring the Shuttle Command

On November 5, 1986, a reorganization of the top command of the National Space Transportation System, the institutional designation for the shuttle, was announced by Dale D. Myers, new NASA deputy administrator. It was designed to centralize control of shuttle operations and program at NASA headquarters in Washington and eliminate the recurrence of the communications gap that, according to testimony, had allowed top shuttle managers to remain unaware of the recommendation of the Thiokol engineers not to launch *Challenger* in cold weather. This was the "flaw" in the launch decision-making process cited by the Rogers commission.

The essential change from the preaccident shuttle

---

[8] The first four shuttle missions were test flights although three carried scientific experiments including a Defense Department experiment.

command structure was transfer of control of shuttle operations and program from the Johnson Space Center, Houston to headquarters, Washington. This was done by appointing Arnold D. Aldrich, shuttle manager at Johnson, as director of the National Space Transportation System and moving his office to Washington. Aldrich was invested with "full responsibility and authority for the operation and conduct of the NSTS program."

Aldrich was the principal figure in the launch chain of command who had survived the wave of executive retirements and resignations that followed the wake of the accident and the commission investigation. He had testified that he had never been informed of the Thiokol engineers' no-go recommendation before the launch. Rear Admiral Truly, to whom Aldrich would report, told a news conference that Aldrich's experience made him "the man for his job at this time."

Two deputies were appointed to assist Aldrich. His deputy for operations was Crippen who divided his time between the Johnson and Kennedy Space Centers. His deputy for program management was Richard H. Kohrs, his former deputy at Johnson. Kohrs would remain in Houston but report directly to Aldrich in Washington. At the Marshall Space Flight Center, the shuttle projects office manager (William R. Marshall) would report directly to Kohrs.

As operations director, Crippen would be responsible for shuttle preparation, mission execution, and the return of the orbiter from California where the shuttle would land for the indefinite future to Kennedy for flight processing. He would be the presenter at the flight readiness review, which would be conducted by Truly. Of first importance, Crippen would take charge of the final launch-decision process. He would also chair the mission management team.

Establishment of the shuttle directorate at headquarters moved the launch and flight decision-making processes out of both Marshall and Johnson. The Kennedy space Center, without a program function, retained some input to shuttle programming as a member of the Space Flight Management Council. This body, mainly advisory, included Marshall, Johnson, and the National Space Technology Laboratories (NSTL), the shuttle main engine test center at Bay St. Louis, Mississippi.

The exclusion of astronauts from shuttle management that had characterized the preaccident managerial system at headquarters was now reversed. With Truly and Crippen in charge of shuttle operations, astronauts would direct the launch decision process for the first time in the history of NASA.

The influx of military and ex-military officers into the management of the civilian space agency evoked comment from a number of observers that it was falling under control of the Department of Defense.

Still, NASA's military and civilian executives insisted that their presence and the increasing military dependence on the shuttle did not signify that NASA was becoming militarized. At a Kennedy Space Center news conference October 2, 1986, McCartney asserted that the boss to whom he owed allegiance was James Fletcher, NASA administrator.

Hard Times

Meanwhile, the mass layoffs of space workers that had followed the end of Apollo also followed the stand down in shuttle flights. At the Kennedy Space Center, approximately 1,100 workers were laid off in February and when it became known that the shuttle would not fly again until the first quarter of 1988, another 1,108 persons were laid off September 18. Most of those laid off were employed by the Lockheed Space Operations Company, which was the unified shuttle processing contractor. Other contractors cutting their work forces at Kennedy were Boeing Aerospace Operations, McDonnell Douglas Corporation, and the Planning Research Corporation. Half of the layoffs reported in February reflected completion of the refurbishment of complex 39B for the launch of *Challenger* 51-L. As mentioned earlier, all previous shuttle launches had lifted off pad 39A. Pad B had been idle since the end of Apollo.[9] The

[9] Both pads were to be used in the spring for the Galileo and Ulysses launches to Jupiter.

autumn work force reductions, however, were the result of the grounding of the fleet. In addition to Kennedy center layoffs, about 800 Martin Marietta employees were laid off at the Michoud, Louisiana external propellant tank assembly plant. At Vandenberg Air Force Base, California, the NASA/Air Force Pacific launch site for the shuttle, about 1,000 Lockheed processing workers were laid off. Vandenberg was to stand down until 1992.

The 1986 employment cutbacks at Kennedy were considered temporary inasmuch as management would have to rehire in a year. Still the retrenchment was reminiscent of the termination of Project Apollo in the 1970s when employment was cut from a peak of 25,800 in 1967 to 9,300 in 1974. By 1981, as the shuttle began flying, the labor force had been rebuilt to 14,000, and by the end of 1985, stood at 16,000.

The ups and downs of employment at Kennedy followed conventional aerospace industry practice, but the managers realized that it was counterproductive in the context of a continuing space program. It took years to develop efficient teams of trained men and women to perform the complex tasks of maintaining, inspecting, processing, and launching the shuttle. The layoffs broke up these teams. It was difficult if not impossible to reconstitute them when space worked resumed. Although some provision for ensuring employment in order to keep a skilled labor force intact had often been discussed, the idea seemed to be foreign to conventional industrial practice. Consequently, the tradition of hire 'em when you need 'em and fire 'em when you don't was the same in space work, where government policy dictated employment levels, as it was in autos or coffee makers, where employment responded to the market.

The Congressional Probe

Four days after the Rogers commission submitted its report to President Reagan, the U.S. House of Representatives Committee on Science and Technology be-

gan a series of formal hearings on the *Challenger* accident, calling 60 witnesses. This committee is the primary overseer of NASA in the House. Among its duties were authorizing funds and reviewing the development of the shuttle. Like its counterpart in the Senate, the Committee on Commerce, Science, and Transportation, the House committee was unaware of the menacing solid rocket motor joint seal problems until they were disclosed by the accident investigation.

The committee's investigation included 10 formal hearings and review of the Roger commission's report and pertinent NASA documents. Its findings in some respects were even more more critical than those of the Rogers commission, although for the most part, they paralleled those of the commission.

In one sentence the committee wiped out what was left of the legend of NASA's technological preeminence, with the statement that "The committee is not assured that NASA had adequate technical and scientific expertise to conduct the space shuttle program properly."

The committee report, made public early in October 1986, pointed out that NASA had been losing top scientific and technical people because it did not match industry salaries. As an example of lack of expertise, it cited NASA's failure to act promptly to remedy joint design flaws when they were reviewed at a conference of Marshall and Thiokol engineers at NASA headquarters on August 19, 1985. "The fact that NASA did not take stronger action to resolve this problem indicates that its top technical staff did not fully accept or understand the seriousness of the joint problem," the committee said.

However, despite its criticism the committee stood back and took a more conciliatory long view of the space agency and its time of trouble.

"We are at a watershed in NASA's history and the nation's space program," it said. "NASA's 28 year existence represents its infancy. We must use the knowledge and experience from this time to insure a strong future for NASA and the U.S. space program in the 21st century."

The committee found that NASA's drive to achieve

a launch schedule of 24 flights a year created pressure throughout the agency that directly contributed to unsafe launch operations. "The committee believes that the pressure to push for an unrealistic number of flights continues to exist in some sectors of NASA and jeopardized the promotion of a 'safety first' attitude throughout the shuttle program.

"The committee, the congress and the administration have played a contributing role in creating this pressure. Congressional and administration policy and posture indicated that a reliable flight schedule with internationally competitive flight costs was a near term objective.

"Pressures within NASA to attempt to evolve from an R & D [research and development] agency to a quasi-competitive business operation caused a realignment of priorities in the direction of productivity at the cost of safety," the report said.

In the main, the House committee concurred with the findings of the Rogers commission, with one notable exception. The House report expressed full agreement on the cause of the accident: joint failure due to a faulty design. The committee report concurred in the commission's conclusion that neither NASA nor Thiokol fully understood the operation of joint before the accident. Further, said the report, the joint test and certification programs were inadequate, and neither NASA nor Thiokol responded adequately to obvious warning signs that joint design was defective.

The committee expressed disagreement with the commission finding that NASA's decision-making process, leading to the launch of 51-L, was flawed. "The committee feels that the underlying problem which led to the Challenger accident was not poor communication or inadequate procedures as implied by the Rogers Commission conclusion," the report stated. "Rather the fundamental problem was poor technical decision making over a period of several years by top NASA and contractor personnel who failed to act decisively to solve the increasingly serious anomalies in the solid rocket booster joints."

Specifically, the report asserted: "The decision to launch 51-L was based on a faulty engineering analysis of the solid rocket motor field joint behavior."

The report went on to say that while Marshall management used poor judgment in not informing higher management levels of the concerns voiced by the Thiokol engineers the night before the launch, "the committee finds no evidence that the outcome would have been any different had they been told."

The House committee stated that it had more concerns than those expressed by the Rogers commission about the safety of orbiter main engine. The committee said that this advanced hydrogen-oxygen engine "may have inadequate safety margins, as indicated by persistent operating problems such as cracked turbine blades, defective hydraulic actuators and temperature sensors." For this reason, the engine should not be operated at a thrust level higher than 104 percent of the design power level. For exceptionally heavy payloads, NASA had planned to increase power to 109 percent. The committee warned against this, stating that payload weight should be reduced instead, as a safety measure, so that power levels could be held at 104 percent or less.

On the subject of communication, which the commission had found inadequate, the House committee took a different view. There was too much of it, the committee said. The internal communication system was disseminating too much information with no discrimination as to its importance, so that "recipients have difficulty separating the wheat from the chaff."

The focus of criticism in the report was NASA's failure to understand how or why deficiencies in the testing certification of the solid rocket motor were not detected. "Without such understanding," the committee said, "NASA will not be able to protect against a similar breakdown in the system of checks and balances in the future."

The committee indicated that it was not happy with the results of consolidating 15 separated shuttle processing contracts into a single contract covering all ground processing for launch and landing. This was done in 1983 under a single contract awarded to Lockheed, the low bidder.

While performance had improved under the contract, it continued to be plagued by excessive overtime, persistent failures to follow prescribed work procedures and inadequate logistical support from NASA, the report said. The committee found that high rates of overtime had increased significantly during the six months before the accident, "to the point that critical personnel were working weeks of consecutive work days and multiple strings of 11 and 12 hour days." This regimen induced worker fatigue that contributed to the accidental dumping of liquid oxygen shortly before the launch of *Columbia* 61-C, the reported noted, "a mission threatening incident."

Moreover, failure of contractor employees to comply with guidelines "in numerous documented cases" had contributed to major mishaps in 1985 during shuttle processing, including damage to a solid motor segment and damage to a payload door.

The space shuttle, the report concluded, "has not yet reached a level of maturity which could be called operational as that term is used in either the airline industry or the military." The report characterized the shuttle as a developmental system.[10]

The committee agreed with the commission that there was no evidence of sabotage, terrorism or foreign covert action in the loss of *Challenger*. It joined the commission and the astronauts in demanding improvements in brakes and landing gear on the orbiter. The new NASA Office of Safety, Reliability, and Quality Assurance must conduct an independent assessment of booster field joint redesign efforts, the report said.

The committee agreed with the commission that launch escape in the present shuttle while the boosters are burning was impossible, but opined that it may be possible to decrease risk to the crew during ascent abort after booster separation.

[10] The shuttle was declared "operational" by NASA on July 4, 1982 with the landing at Edwards Air Force Base, California of *Columbia's* fourth and final research and development flight. The first "operational" flight was *Columbia* STS-5, November 11–16, 1982, when two commercial satellites were deployed.

The Leak Is Reproduced

On the morning of October 27, 1986, the redesign of the booster field joints, nozzle joints, and nozzle that the redesign team had adopted, subject to testing, was presented by John Thomas at a news conference at the Marshall Space Flight Center. It was the product of nine months of data collection and analysis, of the massive salvage operation in the Atlantic Ocean, of hours of testimony before the Presidential Commission, thousands of feet of video film of the launch and holocaust, hundreds of engineering documents, and months of conferences.

Essentially, the redesign fixed the original booster joint design at its points of weakness by adding the capture device, improving the O-ring material, reducing gap tolerance, replacing putty with insulation, and installing heating strips. But the climax of the billion-dollar effort to refly the shuttle did not come until fall, when conditions that had caused the disastrous *Challenger* booster leak were reproduced in a test motor and promptly reproduced the leak.

The tests were run at 25° F in a joint environmental simulator at Morton Thiokol on two full motor segments assembled with two *Challenger* 51-L–style joints. They showed that a joint without putty leaked from the start of the ignition transient when the old-style O-rings were used, but when improved O-rings were installed, only the primary O-ring leaked while the secondary ring held the gas pressure.

In a 51-L–type joint where putty was used, leakage did not occur until a blowhole was made in the putty, such a hole as would be made by a leak check. Then, Thomas related, "we had smoke through the joint at the point where the blowhole was placed. And the smoke appeared to be visually about the same as we saw on 51-L. It leaked for the full duration of the test."

The redesign chief said that the test series accomplished two results. "We have determined that our failure investigation results are in fact valid and that

FIELD JOINT METAL AND INSULATION

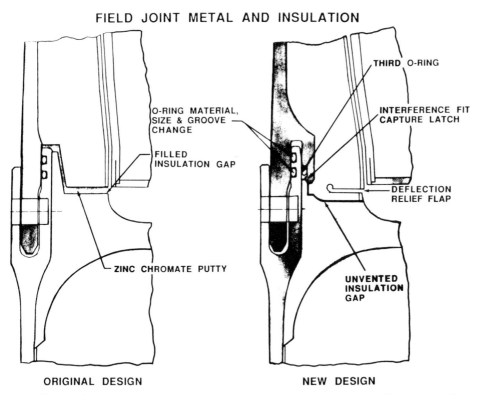

Basically, the fix in the solid rocket motor field joints added a capture latch and a third O-ring, as shown here. (NASA)

NOZZLE/CASE JOINT METAL AND INSULATION

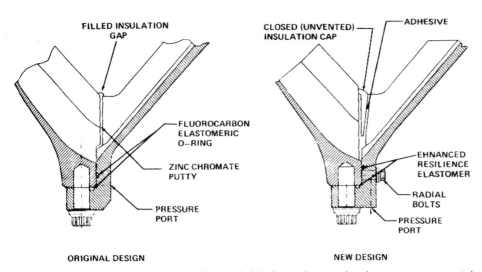

The redesigned nozzle-case joint added radial bolts, a feature that became controversial. (NASA)

we're proceeding on a course to change those factors . . . that produced the 51-L accident.''

Thomas was asked: ''Are you satisfied now that you have reproduced the 51-L leak scenario?''

''Yes,'' he said. ''We've reproduced every aspect of the condition and the results. It's the first time that has been done in full scale. Fundamentally, what this did was confirm our earlier conclusions of what happened and also confirmed that the design solutions that we have under way will in fact take care of the problem we did experience. It confirmed to us that we're headed in the right direction.''

In the event continued testing discloses flaws in the redesign, the team has considered alternate designs from the new ''baseline.'' One would provide a vent in the joint insulation through which gas could escape to relieve pressure if the tests showed it stressed insulation bonding. The venting would create a labyrinthine passage through which gas could flow between ablative surfaces. The function of the passage is to allow hot gas to cool sufficiently so that when it reaches the O-rings it is no longer hot enough to damage them. Tests have shown that venting the joint insulation appears effective in relieving pressure.

Thomas reported that the redesign team was responding to points raised by the National Research Council panel concerning additional full-scale motor firing tests. NASA had agreed to build a second full-scale horizontal test facility at Thiokol, and one of two additional tests the panel recommended would be run on this test stand. Thomas said the additional tests would not affect the projected February 1988 launch date.

The additional test firings were not prompted by the council panel's recommendations, Thomas said. They had been considered previously.

''We recognized at the very outset that we had a minimal test program, although sufficient to get us through the redesigns that we had in mind at that time,'' he said. ''We're right now looking at ways of increasing the number of tests in the joint environmental simulator and

methods and mechanisms for accomplishing the two additional tests.''

By Mutual Agreement

The booster design team evolved rapidly into an institutional force, made up of NASA and contractor engineers across the spectrum of solid rocket propulsion manufacturing. The NASA component, in addition to Marshall people, included specialists from the Langley Research Center in Virginia, the Johnson Space Center in Texas, the Lewis Research Center in Ohio, and the Jet Propulsion Laboratory in California. There was representation of the team from the Kennedy Space Center in Florida and the Corps of Astronauts.

Thiokol engineers and their associate contractors and subcontractors participated fully, Thomas said. ''We stay very close together in our discussion of the design and in fact we don't proceed with any design feature unless we have mutual agreement that that is the proper way to go.''

Final confirmation of the redesign team's work was scheduled to be decided at the critical design review in May 1987, according to the plan projected at the end of October 1986. If the redesign was confirmed, the first flight motors would be ready for assembly at Kennedy in October 1987.

''We see no reason now to deviate from our plan of February 1988,'' Thomas said. ''We are confident that the test program will confirm that the design we have right now is the design we are going to fly.''

Constrained by inventory and a noncritical target date, the redesign team was carrying a heavy load. Within these limitations, its burden was to produce a joint that would never fail. It was on this basis that shuttle flight would be resumed without a crew escape system. As the Rogers commissions had stated: ''The present shuttle has no means for crew escape either during the first stage ascent [boosters firing] or during

Descending on its three ringsail parachutes, Apollo 14 splashes down safely in the Pacific Ocean southeast of American Samoa on its return from the Moon, February 9, 1971. Aboard were Astronauts Alan B. Shepard, Jr., Stuart A. Roosa, and Edgar D. Mitchell. At splashdown, the Apollo command module weighed 11,500 pounds. (NASA) The Rogers commission noted that the orbiter crew module weighed 14,000 to 17,000 pounds. An Apollo-style parachute system may one day provide launch escape capability for the orbiter crew module.

gliding flight. The crew must remain in the orbiter and its only option is to attempt emergency landing which is possible only after solid rocket booster burnout.''

In 1971, Rockwell International had considered three launch escape modes: ejection seats, encapsulated ejection seats and a separable crew compartment. Compared to a $10 million ejection seat weighing 1,760 pounds, the separable crew module would weigh 7 to 8½ tons and cost $292 million (in 1971 dollars), the commission reported.

The commission said that conventional ejection seats do not appear to be a viable option because they limit crew size and thus restrict shuttle missions. Other options examined were the separable crew compartment or escape module that would be detached from the orbiter and descend by parachute; rocket-assisted extraction from the crew compartment using small rock-ets to boost occupant and parachute out and away from the orbiter; and a bail-out system enabling crew persons to make an unassisted exit through a hatch during gliding flight and descend by parachute.

The bail-out system was recommended by Admiral Truly, and Administrator Fletcher confirmed its adoption when he visited the Kennedy Space Center on a tour of inspection on February 12, 1987. This system required the installation of a hatch that could be jettisoned quickly—a fix adopted for the Apollo command module after the Apollo fire of 1967. It would be useful, Fletcher said, only for a relatively narrow segment of the flight envelope, during gliding flight at medium altitudes and over water. Each crew member would jump out with parachute and rubber boat.

The administrator said that the fast-jettison hatch would be available for the resumption of shuttle flight

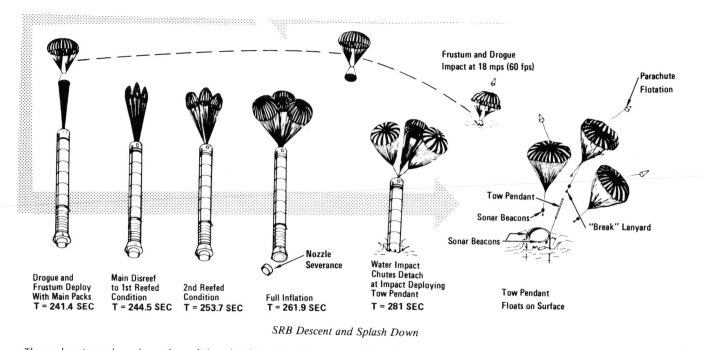

| Drogue and Frustum Deploy With Main Packs T = 241.4 SEC | Main Disreef to 1st Reefed Condition T = 244.5 SEC | 2nd Reefed Condition T = 253.7 SEC | Full Inflation T = 261.9 SEC | Water Impact Chutes Detach at Impact Deploying Tow Pendant T = 281 SEC | Tow Pendant Floats on Surface |

SRB Descent and Splash Down

These drawings show how the solid rocket booster with a dry weight of 180,000 pounds descends on three 115-foot conical ribbon parachutes. Velocity at water impact is 87 feet per second (53.79 miles per hour). (Rockwell)

with the launch of *Discovery* targeted for February 18, 1988, but he added that he did not know whether it would be actually implemented. Uncertainty and skepticism about the survival value of this escape system had pervaded the decision to try it. As testimony before the Rogers commission had shown, there was concern that a crew member jumping out of the hatch might not clear the left wing.

A tractor rocket device that would pull the astronaut clear of the orbiter had been considered but was deemed too risky to try without further study. Fletcher said that the quick-jettison hatch had both "pluses and minuses."

To observers it was clear that bail-out was simply a concession to concern expressed by members of the Rogers commission that some sort of escape system should be provided and that bail-out would cause the least delay and excess cost. At least, it appeared to be better than doing nothing. No one at NASA headquarters expressed any enthusiasm for it.

The separable crew compartment offered "the widest range of crew escape options" in theory, the commission report said. It would provide crew escape at all altitudes during a first stage (boosters firing), time critical

emergency if the escape system itself is not damaged.

However, the report went on, the escape module would require "significant redesign of the orbiter, some structural reinforcement, pyrotechnic devices to separate the escape module from the rest of the orbiter, modifications to sever connections that supply power and fluids, separation rockets and a parachute system."

Could an escape module survive the holocaust that destroyed *Challenger*? A number of experts—astronauts and engineers—did not think so. Still, the photographic record indicated that survivability was possible. The crew compartment was separated intact from the orbiter by explosive forces and was photographed emerging intact from the fireball, falling nose forward toward the ocean.

As the Kerwin report indicated, the crew may have survived these forces and possibly the decompression that followed, but perished when the compartment struck the ocean. Whether an escape module with parachutes could have saved them remains moot. How can it be dismissed?

As the commission stated: "The space shuttle system was not designed to survive a failure of the solid

End of mission, April 29, 1986. Remains of the *Challenger* 51-L crew depart the Kennedy Space Center. (NASA)

rocket boosters.'' Such was the basic risk designed into the shuttle—the only manned spacecraft without provision for launch escape. The omission was a design decision. The *Challenger* accident has produced little impetus to alter it. In listing escape options, the commission made no recommendation.

Shuttles will fly again under these circumstances, with improved booster joints.[11] Crews thus should be safer than before. But they will not be launched fail safe, for they will still be committed to the risk that the crew of *Challenger* 51-L did not survive.

[11] In lieu of the $10 million penalty imposed by its NASA contract for catastrophic failure of a solid rocket motor, Morton Thiokol agreed to waive $10 million in profit from the motor program, according to an understanding announced by NASA on February 24, 1987. In addition, the company agreed to waive profit on $409 million in manufacturing required to redesign faulty field joints, rework existing hardware to fit the new design, and replace reusable motor hardware lost in the *Challenger* accident. NASA said that a modification of the $1.3 billion motor contract reflecting this understanding would be signed by October 1987.

Epilogue: Mission 26

Early in January 1987, NASA scheduled a new beginning for the shuttle. The booster fields and nozzle joints were being redesigned and the orbiter engines were undergoing intensive reviews. Rear Admiral Richard H. Truly, Associate Administrator for Space Flight at NASA headquarters, announced a flight crew of five members for the launch of the space shuttle *Discovery* on February 18, 1988. The flight was designated simply as Mission 26. It would last four days and land at Edwards Air Force Base, California.

Preparations for Mission 26 were carried on at the Kennedy Space Center with perseverative attention to detail. It was said that inspections were made and reinspected as if the orbiter were being processed for the first time. A morbid concern about oversight and error permeated the NASA organization from headquarters through the centers. It was fed by reports of criticism from members of scientific and technical advisory groups who contended the NASA was not going far enough to correct deficiencies that the Rogers commission had cited.

The total effect of this mood was to displace the confidence and optimism that had characterized NASA flight operations since Project Mercury. Space flight had become a grim business, and its resumption was a dire challenge to the American aerospace establishment. The stakes were high because of the widespread belief in much of the space community that another shuttle disaster would ground the National Space Transportation System and with it the space station for a decade or more. *Discovery* could not be allowed to fail. And that put great pressure on every person who had a role in getting it ready for flight.

Navy Captain Frederick H. Hauck, 45, was named *Discovery* mission commander. He had been assigned months before as Acting Associate Administrator for External Relations at NASA headquarters. Now, he returned to astronaut duty. Hauck had commanded *Discovery* mission 51-A, November 8–16, 1984 and had been pilot aboard *Challenger*'s STS-7 flight, June 18–24, 1983.

The pilot assigned to Mission 26 was Lieutenant Colonel Richard O. Covey, 40, of the Air Force. He had flown as pilot on the 51-I *Discovery* mission of August 27–September 3, 1985. Three other experienced astronauts were assigned as mission specialists. They were John M. Lounge, 40, an astrophysicist and also a human factors specialist; Dr. George D. Nelson, an astronomer; and Marine Corps Major David C. Hilmers, 36, an electrical engineer. *Discovery*'s primary payload was to be TDRS-2, second of three planned tracking and data relay satellites in NASA's new space-to-ground communications system. It would replace the satellite lost on Challenger.

Truly approved the installation of an explosive hatch for the shuttle crew compartment, and technicians at KSC prepared to install it in *Discovery*. It would enable crew persons to make a quick exit from the orbiter in an emergency, equipped with a personal parachute, inflatable life raft, and signaling device. This escape system was survivable only after the solid rocket boosters had burned out, had been dropped, and the orbiter was in a stable glide over water. Addition of an extraction device consisting of a small rocket that would pull a jumper away from wing or tail surfaces had been considered but was dropped from the escape system.

By the first quarter of 1987, the hatch modification and other alterations made it apparent that the February 18 launch date could not be met. KSC people began to predict a launch no earlier than the second quarter of the year.

Early in April 1987, the scientific observatories, Galileo and Ulysses, that seemingly had been consigned to limbo after the *Challenger* accident, were rescheduled for shuttle flights in 1989 and 1990. Both had been scheduled for launch in May 1986.

NASA announced that Galileo would be launched in early November 1989 on a six-year flight to Jupiter. It would go into orbit to observe the planet and its major moons and drop a probe into the atmosphere. The following year, the European Space Agency's solar observatory, Ulysses, would be launched on a similar path. However, it would pass the big planet on a trajectory allowing Jovian gravitation to hurl it into a polar orbit around the sun.

A remarkable feature of these missions was their unique trajectories, which had been designed by NASA's Jet Propulsion Laboratory to utilize gravitational assist to reach Jupiter. This was necessary to compensate for a reduction in rocket propulsion.

Initially, the observatories were to be boosted directly to Jupiter by hydrogen-fueled Centaur rockets after being deployed in low Earth orbit by their respective shuttles, *Challenger* and *Atlantis*. Both orbiters had been modified to carry the volatile Centaurs. Postaccident safety concerns, however, persuaded NASA to substitute the less volatile but also less powerful Inertial Upper Stage boosters that used solid propellants.

The switch would more than double the flight time by requiring a detour around Venus for gravitational assist. This involved an unprecedented celestial navigation program.

After deployment in low Earth orbit, the observatories would be boosted on a flight path around Venus that would bring them back to Earth. They would be accelerated by Earth's gravitation around the sun and once more return to Earth, this time on an outbound trajectory. They would pass Earth closely this time, at an altitude of about 186 miles, grazing the top of the atmosphere.

The second Earth passage would occur about three years after launch. It would impart enough energy to Galileo and Ulysses to hurl them away from the sun, out to the orbit of Jupiter.

Galileo's onboard propulsion system would drop the spacecraft into Jovian orbit. Ulysses, however, would be accelerated by Jupiter above the ecliptic plane into a solar orbit that would pass the as yet unseen poles of the sun.

It had become apparent to KSC workers early in the year that new safety requirements, processing procedures, and pressure for additional testing would scrub the February 18, 1988 launch date. On April 13, Truly announced a decision to conduct two prelaunch tests while *Discovery* was on the pad. One would be a wet countdown demonstration test with the shuttle external tank filled with propellant. The second would be a flight readiness firing of the main engines for 20 seconds. Such tests usually were done before the launch of a new vehicle.

Truly's announcement confirmed forecasts by space workers that the tests would add several weeks to the processing time for *Discovery*. Arnold D. Aldrich, director of the National Space Transportation System office at headquarters, advised that *Discovery* could not be launched before April 1. KSC engineers predicted a longer delay. The launch date ultimately depended on the outcome of solid rocket motor tests during the summer. The National Research Council had recommended two full-scale booster firing tests in addition to the tests planned, but NASA had not accepted the recommendation.

During this period of tension and uncertainty, a new American space transport appeared on the horizon

GALILEO OCTOBER - NOVEMBER '89 VEEGA

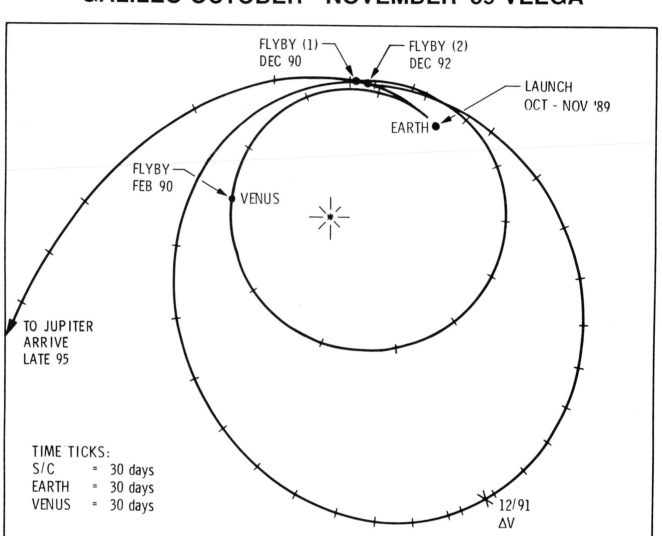

Because of its unique gravity-assist features involving both Earth and Venus, the Galileo mission to Jupiter is to fly a trajectory called VEEGA (Venus-Earth-Earth Gravity Assist). Based on the planned VEEGA launch, November 4, 1989, Galileo would fly by the sunward side of Venus on February 19, 1990 at an altitude of 12,428 miles above the surface. Venus would impart a velocity increase of 1.4 miles a second to the spacecraft (relative to the sun) and direct it back to Earth. Galileo would pass Earth at an altitude of 2,237 miles on December 12, 1990 and acquire additional velocity of 3.24 miles a second. The increase would raise its aphelion (maximum distance from the sun) to the asteroid belt. An onboard propulsion maneuver of 82 meters a second would align Galileo for a second Earth passage on December 6, 1992, this time at an altitude of 186 miles from the surface on the sunlit side. Earth's gravity would add 2.3 miles a second to Galileo's velocity—enough to send the spacecraft to Jupiter by November 29, 1995. Ulysses would follow a similar trajectory. (JPL-NASA chart)

ORBITER

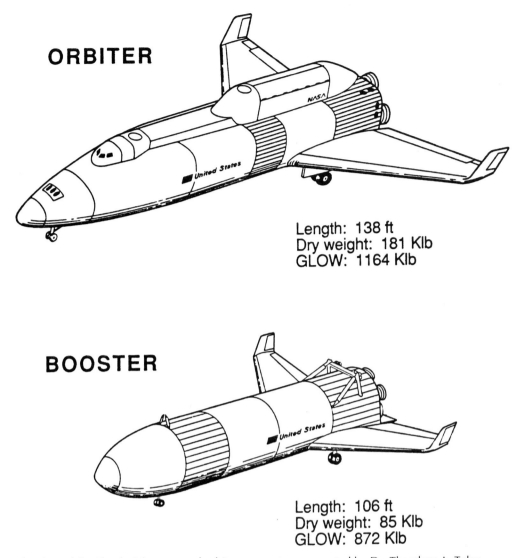

Length: 138 ft
Dry weight: 181 Klb
GLOW: 1164 Klb

BOOSTER

Length: 106 ft
Dry weight: 85 Klb
GLOW: 872 Klb

Sketches of the Shuttle 2 booster and orbiter concepts as presented by Dr. Theodore A. Talay of the Langley Research Center to the Twenty-Fourth Space Congress of the Canaveral Council of Technical Societies, April 23, 1987, at Cocoa Beach, Florida. The legend GLOW refers to gross liftoff weight, which is expressed in thousands of pounds. (Courtesy, Langley Research Center)

of the twenty-first century—Shuttle 2. In its initial design, it was smaller and could be launched more quickly than the current shuttle. It had neither solid rocket boosters nor external propellant tank.

Shuttle 2 had been in the process of definition at NASA's Langley Research Center since 1985. It would carry a crew of two to five, fly missions up to three days' duration, and service the space station and unmanned orbital platforms. It could be processed and

launched every two weeks and require only 24 hours on the pad.

Shuttle 2 was described in a progress report by Dr. Theodore A. Talay of Langley to the Twenty-Fourth Space Congress of the Canaveral Council of Technical Societies on April 23 at Cocoa Beach, Florida. Shuttle 2 would consist of a flyback booster and an orbiter with internal propellant tanks. Both had been considered and dropped in 1972. To contain the internal propellant tanks, the orbiter would be 138 feet long, 16 feet longer than the present orbiter. Internal tankage does not limit payload bay size or weight, Talay said. The payload would be contained in a canister that is placed on top of the fuselage. The aerospace technologist described a payload container 30 feet long (half the length of the present cargo bay) and 15 feet in diameter (the same as the existing bay). It would have 20,000 pounds capacity, about one-third present capacity. However, the flatbed placement of the payload canister on the fuselage would allow for a longer container if required, Talay said.[1] Talay explained that heavy lift requirements could be met by shuttle-derived expendable boosters and later by partially and fully reusable launch vehicles.

The orbiter, powered by five reusable hydrogen-oxygen engines of 300,000 pounds of thrust each, would be launched vertically astride its 106-foot, winged booster powered by six reusable engines of 250,000 pounds of thrust each, burning methane and liquid oxygen. All 11 engines would fire at lift-off with a total thrust of three million pounds.

The booster would carry three propellant tanks containing liquid oxygen, liquid hydrogen, and the hydrocarbon fuel for the booster engines. The hydrogen and oxygen would be cross-fed to the orbiter during the first stage of launch. This system would allow the orbiter tanks to remain filled until the orbiter separated from the booster at Mach 3 (90,000 feet). Detached from the orbiter, the booster would glide back to an automated landing at the launch site runway.

[1] Theodore A. Talay, personal communication.

Shuttle 2 would have automated, self-diagnostic checkout systems with built in test equipment. By making it as autonomous as possible with expert systems, robotics, and artificial intelligence, labor-intensive operations including many mission control functions could be reduced, Talay said.

Safety is a major design concern, he said. Shuttle 2 would have an escape system to enable the crew to survive an accident under "worst case conditions."

The forward flight compartment would be rocketed free (an 8-g maneuver) by solid rocket motors under conditions where crew survival is threatened. The escape system would employ module stabilization devices and parachutes.

Landing bags and possibly terminal descent motors would null out the vertical landing velocity, said Talay.[2]

As vehicles in the original shuttle fleet wore out, the new generation of shuttles would replace them. This would occur early in the 2000 time period.

Meanwhile, the first steps toward the launch of Mission 26 in September 1988 were taken in the summer of 1987. At the Kennedy Space Center, *Discovery* was powered up for the first time in two years for a test of its electrical systems on August 3. This was followed on August 30 by the first full-duration (2 minutes) firing test of a redesigned solid rocket booster at the Morton Thiokol horizontal test stand, Promontory, Utah.

More tests were to follow as the space agency moved cautiously to restore manned space flight in America for the balance of the twentieth century.

[2] Personal communication.

Index